Wheels and Axles

Organizing Committee

K Rose (Chairman)
Railtrack plc

D Anand
Toton Sidings

J Allen
(Mechanical) Trains Delivery Group, LUL

J Benyon
Traction and Rolling Stock, AEAT Rail

Papers presented at a one-day seminar *Wheels and Axles* held at IMechE Headquarters, London, UK, on 21 April 1999.

IMechE
Seminar Publication

Wheels and Axles

Organized by
Railway Division of
The Institution of Mechanical Engineers (IMechE)

Co-sponsored by
The Institution of Railway Signal Engineers
and The Institution of Civil Engineers

IMechE Seminar Publication 1999–12

Professional Engineering Publishing

Published by Professional Engineering Publishing Limited for the Institution of Mechanical Engineers, Bury St Edmunds and London, UK.

First Published 1999

ISSN 1357–9193
ISBN 1 86058 242 7

A CIP catalogue record for this book is available from the British Library.

Printed and bound in Great Britain by Antony Rowe Limited, Chippenham, Wiltshire, UK.

Related Titles of Interest

Title	**Editor/Author**	**ISBN**
IMechE Engineers' Data Book	Clifford Matthews	1 86058 175 7
New Trains for New Railways	IMechE Conference	1 86058 146 3
Developing and Growing the Business – Passenger	IMechE Seminar	1 86058 182 X
Developing and Growing the Business – Freight	IMechE Seminar	1 86058 183 8
Getting Trains into Service	IMechE Seminar	1 86058 186 2
Technology for Business Needs	IMechE Seminar	1 86058 185 4
Railway Traction and Braking	IMechE Seminar	1 86058 018 1

For the full range of titles published by Professional Engineering Publishing contact:

Sales Department
Professional Engineering Publishing Limited
Northgate Avenue
Bury St Edmunds
Suffolk
IP32 6BW
UK

Tel: +44 (0)1284 724384
Fax: +44 (0)1284 718692

Contents

S629/001/99

Rickerscote – three years on

K A ROSE
Railtrack plc, London, UK

INTRODUCTION

In March 1996 an accident occurred at Rickerscote on the West Coast Main Line near Stafford caused by the failure of an axle on a freight wagon. The ensuing derailment of the freight train unfortunately resulted in a Royal Mail train travelling in the opposite direction colliding with it, and the tragic loss of one life on the latter train. The cause of the accident was a fatigue failure of the mid axle span of the freight wagon, and the HMRI inquiry into the accident concentrated on the causes and prevention of such an event in future.

The HMRI inquiry report was issued in November 1996 and contained eleven recommendations. Seven of these were specifically to do with axle fatigue failures, and their detection and prevention. The HMRI required Railtrack to co-ordinate the response to these inquiry recommendations and as a consequence in January 1997 what was to be called the Rickerscote Steering Group was set up to take forward specific recommendations. This Steering Group was chaired by Railtrack Safety & Standards Directorate and consisted of representatives from all of the owners of axles operating on Railtrack Infrastructure. There were thus representatives from the Rolling Stock Leasing companies, Association of Train Operating Companies, freight operating companies and the Private Wagon Federation.

The Steering Group took the seven recommendations of the HMRI and merged them into the four tasks below:

1. Review of the mechanisms of axle failure through crack propagation and methods of reducing their incidence.
2. Investigate the reasonableness of changing the current axle testing regimes.
3. Investigate the value of conducting a metallurgical examination of rejected axles.
4. Investigate the benefits and practicalities of collating and analysing axle rejection statistics.

To carry out these tasks the Steering Group sought tenders from industry and in June 1997 a contract was let with AEA Technology for this work. The Steering Group was keen to ensure that the work concentrated on practical solutions rather than any theoretical research, and also that due consideration was taken of experience of other similar railways, for instance, those in Western Europe and in North America. The work was completed in early 1998, and the conclusions of the work were agreed jointly with the HMRI in February 1998. Copies of the final report were sent to all Train Operating Companies.

AEA STUDY

The review of axle failure rates in the UK compared with other administrations' practice demonstrated that in other countries the axle breakage rates were considerably lower than in the UK. However upon investigating the design criteria used elsewhere there was no indication that higher design stresses were being used in the UK. Similarly the level of corrosion protection used on UK axles and the ability to predict fatigue crack growth rates was found to be at least as good in the UK as elsewhere. Certain differences that were noted however included the more general use elsewhere of standard axle designs, such as AAR and UIC axles. The normal practice in the UK has been to design specific axles for given applications rather than using as it were a standard axle from a catalogue of agreed designs. One other issue of difference noted was that past practice in the UK has been to use quenched and tempered heat treatment of axles as opposed to normalised material. The view of metallurgists was that was unlikely to be significant.

The load input into an axle is a function of the track quality and the wheelset maintenance standards. The AEA study showed that measurements in service on UK axles had demonstrated exceedance of design stress when running over poor track and also the influence of large wheelflats on increasing stresses above design levels. Although objective evidence to allow comparison with other railways was difficult to obtain in this area, the indication was that the track quality in the UK could be worse than that experienced by similar axles abroad.

A review of the crack detection methods used in UK rail industry concluded that the key method used for detecting cracks in the main body of the axle between the wheels, ultrasonic "far end" scan, had an unacceptably low probability of detecting cracks. It was concluded that although the operators were adequately trained and the ultrasonic equipment used was acceptable, the technique is just not suited for determining cracks in such a location on the axle. However, the use of ultrasonic inspection techniques for axle journals and wheelseats ("near end" and "high angle" scans) was underwritten by the study which concluded that the probability of detection using these techniques was acceptable.

AEA therefore proposed that alternative inspection techniques should be used for checking the integrity of the main body of the axle, and several alternative techniques were proposed. It was noted that the UK was almost unique in the railway world in having a regime of in-service ultrasonic inspection of the axle body, as almost all of the railway administrations operate a regime of no intermediate NDT inspection of the axles between wheelset overhaul. The final major conclusion of the work was that there was valuable intelligence to be gained from gathering information from axles rejected due to cracking. The data should include such fields as axle serial number, design type, position of defect, location and type of test. It was recommended that this information should be collated centrally to allow improved management of axles using the greater amount of data which is available from across the whole of the UK axle population rather than just the specific fleet information which any given engineer may have at present.

Work since that time by the Steering Group has been aimed at implementing the agreed conclusions from the work and this is described in more detail in the following sections.

MAGNETIC PARTICLE INSPECTION

Although the AEA study recommended the investigation of various new techniques not previously used on the railway, they in addition suggested that as a stop gap measure magnetic particle inspection should be required of all axles in the interim. This action was agreed and it was decided that a Railway Group Standard would be written which required magnetic particle inspection as an essential part of axle maintenance.

The Standard was drafted and requires that all axles undergo either magnetic particle inspection or an equivalently reliable technique at least at wheelset overhaul frequency. The definition of wheelset overhaul frequency has proved somewhat difficult, but the essence of it is that whenever the wheelset was due to receive maintenance off the vehicle then MPI should be carried out. This does not however mean more stripping of components from the axle than would have otherwise been done at that level of overhaul. In addition the Standard contains an appendix giving a best practice guidance on MPI testing covering issues such as the specification and calibration of the test equipment.

During this work it was identified that some of the MPI plants currently in use within the UK were not being calibrated in an appropriate fashion to maximise their potential to detect cracks.

Discussions with the HMRI have taken place over the transition arrangements for introduction of this Standard to existing axles and these discussions have now been satisfactorily resolved. The Railway Group Standard will therefore be issued shortly.

NEW NDT TECHNIQUES

Various new NDT techniques have been identified during the period of this work and are the subject of other papers today. The following is therefore limited to a list of those which have received some attention by the Rickerscote Steering Group.

1. Alternating current field measurement (ACFM), a product of the company TSC.
2. NEWT EMA technology, now being marketed by Alstom.
3. Eddy current phase plane techniques.

AXLE STRESS ENVIRONMENT

Tests will commence shortly to establish if there are any features on the Railtrack infrastructure which are causing higher than the expected stress levels in axles. These tests will use a strain-gauged axle run over a significant portion of the network and will allow correlation with the specific track features. The objective of the work is to identify track features which are critical to axle fatigue life and the population and magnitude of such features on the Railtrack network.

As a result of this work it should be possible to determine where changes may be necessary, either to the design criteria currently used for UK axles, or to track maintenance standards.

NDT DATABASE

It has been clearly recognised that there is benefit in gathering together the NDT test information from axles to allow an industry wide view of the performance of axles. The steering group has investigated various possibilities for achieving this whilst attempting to ensure that any new requirement did not impose onerous needs for multiple system input or use of different systems.

It is most likely that the requirement will be in the form of a Railway Group Standard which dictates that axle NDT inspection data is made available in a common format which can easily be assembled from all of the train operators/maintenance locations into an industry standard database. This database can then be made available to each train operator to assess either individually or in joint industry forum the performance and any subsequent actions necessary for their own regime.

CONCLUSIONS

The main conclusions of the outcome of the Rickerscote inquiry and the work by the Rickerscote Steering Group so far may be summarised as follows.

1. Crack detection in axles which depends on far end ultrasonic scans is unreliable.

2. Many alternative NDT techniques are available which can produce a more reliable detection of cracks than the current far end ultrasonic scan technique.

3. Several of the new techniques are being implemented as the front line crack detection techniques.

4. The formation of a UK database for axle NDT records is justified and will shortly be put into place.

5. Whilst alternative new NDT techniques are developed there will be a Railway Group Standard requirement that all axles undergo Magnetic Particle Inspection as part of their maintenance regime.

6. Tests using an instrumented axle will take place which will allow verification of the design criteria used for axles in the UK.

European standards for wheelsets

J R SNELL
The Engineering Link, Derby, UK

1. INTRODUCTION

The railway industry in the United Kingdom has a long history of regulation, initially by Parliament and the Crown and latterly by Government agencies. Within this environment both industry and National standards have evolved for the selection of components that are key to safety. Arguably the most safety critical of all vehicle components is the wheelset. Consequently in terms of operation on Railtrack lines, there are now 19 standards for wheelsets and their prime components in effect:-

- 6 British Standards
- 13 Railway Group Standards

and these are supported by many other industry standards and procedures either inherited from British Railways or generated subsequently to meet new business needs.

With such an abundance of proven and familiar standards, why now the need for European Standards too?

The reasons lie in the growth of the European Union (EU), with the requirement for public procurement to be on a level playing field throughout the EU, and the development of the concept of railway interoperability across existing frontiers, particularly when coupled with the development of new Trans European high speed routes. The production of standards designed to serve this new environment has given the opportunity to introduce the best of current practice across Europe into a common set of standards and to introduce improvements where technical advance and associated service experience has meant that the existing standards are no longer adequate.

2. THE EUROPEAN REGULATORY FRAMEWORK

The European Union has issued various "New Approach" procurement Directives covering a wide range of sectors. These stipulate that compliance is required with common European - wide technical specifications which define "Essential Requirements."

The purpose of a European Standard is to translate into industrial terms the "Essential Requirements" contained in a "New Approach" Directive. Compliance with the relevant European Standard deems compliance with the "Essential Requirements." This confers authority to place the component or system into service, subject to compliance with conformance assessment procedures.

Two Directives cover the railway sector:-

- Directive 93/38 - Government Procurement Directives (June 14 1993);
- Directive 96/48 - Interoperability of the Trans-European High Speed Rail System (July 23 1996).

2.1 The "Government Procurement Procedures" Directive draws up rules enforceable for awarding contracts in the water, energy, transport and telecommunications sectors. It states that the technical specifications used in contracts shall be based on European Standards, wherever they exist.

2.2 The Directive for the Interoperability of the Trans-European High Speed Rail System lays down requirements for planning, building and operating the new network infrastructure and the requirements for the new rolling stock to be operated over it. It defines the requirements for Technical Specifications for Interoperability (TSI's) which specify necessary components and their interfaces that need to be addressed by European Standards.

2.3 The suite of new European Standards for wheelsets are needed to meet the requirements of both the above Directives. The new Standards will address both "Interoperability" and "New Approach" requirements because they will encompass conformance assessment methods as well as design standards.

3. CEN STRUCTURE AND EN STANDARDS

CEN (The European Committee for Standardisation) is a non-profit making international association for technical and scientific purposes. It is a federation of the National Standards Institutions of 18 countries - the member states of the EU and EFTA (The European Free Trade Association) together with, from 1 January 1997, the Czech Republic. Whilst an independent body, it may work under a mandate from the European Commission, as has already happened with the Operability Mandate No M/275.

CEN's prime remit is to draw up European Standards (EN). Their production is by consensus of the participating bodies and they are confirmed, after a series of structured National

Enquiries, by qualified majority voting. Once agreed, an EN becomes mandatory amongst the participating Nations and any conflicting National Standard must be withdrawn.

CEN has set up over 280 Technical Committees (TC's). Production of EN's is entrusted to Sub Committees (SC's) who either carry out the work themselves, or, where the subject matter is diverse set up their own Working Groups (WG's) to execute the task. Provisional EN's (prEN) are subject to scrutiny at each senior level before National Enquiry and voting, although there are options for the intermediate scrutiny levels to be waived if a prEN is considered to be sufficiently mature.

4. CEN WHEELSETS WORKING GROUP CEN/TC256/SC2/WG11

CEN has assigned rail applications to Technical Committee TC256. Its Sub Committee SC2 has responsibility for bogies and wheelsets. SC2, in turn, has set up six Working Groups and WG11 carries responsibility for wheels and wheelsets.

WG's are populated by recognised experts in their fields nominated by the National Standards Bodies, although they are not subject to mandate by their Standards Bodies. WG members are sponsored by the sectors they represent, not the nominating Standards Bodies.

WG11 was instituted in 1992 and has been supported by the National Railway Administrations of Belgium, Britain, France, Germany, Italy and Sweden and all the wheelset manufacturers of the EU together with, latterly, the Czech Republic. Since privatisation of the railways in Great Britain the UK railway position has been sponsored first by Railtrack and subsequently by the Rolling Stock Companies.

5. EXISTING WHEELSET STANDARDS

In the UK, wheelsets for domestic traffic on Railtrack lines are governed by the 6 parts of British Standard BS5892 as called up by the 13 parts of Railway Group Standard GM/RT2525 (the wheelset manual). In addition wheelsets of vehicles in International traffic may conform to UIC standards. None of these documents alone meets the remit of the EU Directives since most define product delivery state characteristics and procedures, although the wider structure of Railway Group Standards called up from the Wheelsets Manual does address some of the issues relating to product and supplier qualification.

UIC leaflets again address product delivery issues but also they are not appropriate for meeting EU requirements, because their authorship and validation was restricted to member railways, not the supply industry.

ISO standards also exist for railway wheelsets but their applicability to the new European context is limited, because their structure has tended towards acting as a catalogue of alternative solutions rather than the derivation of a definitive, single standard.

6. THE EUROPEAN STANDARDS

To meet the EU Directive aspirations, the new European Wheelset Standards fall into two specific groups:-

- Technical approval Standards;
- Product and supplier qualification and product delivery Standards.

The Technical approval Standards determine specimen baseline duties, fitness for use criteria and their means of assessment.

The Product Standards interface with Technical Standards and describe the characteristics required of products which have met the technical criteria.

The structure follows a consistent format:-

- Definition of characteristics (metallurgical, mechanical, geometric etc.).
- Product qualification requirements, which provide a benchmark for assessing that a supplier has the capability to consistently supply a product which achieves the technical requirements. These comprise:-

  a) a statement by the supplier of his resources, experience and capability to supply the required product;

  b) auditing of the supplier to verify the case stated in a).

  c) testing of the product against technical requirements to validate its claimed status;

  d) a monitored period of supply to verify manufacturing consistency;

- Product delivery (supply) procedures. These give the option of testing of batches or mandatory testing as part of an agreed quality plan.

6.1 Wheelset Technical Standards

Three Technical Standards are now at an advanced state:-

- Wheel Technical Approval
- Power Axle Design Method (prEN13103)
- Non-Power Axle Design Method (prEN13104)

6.1.1 Wheel Technical Approval

The remit of WG11 is to sythenise the best of current European practice to produce a workable standard. It must consider any existing European National Standards as part of this process, followed by UIC Standards. As referred to above, ISO Standards have not been seen to add to this process. WG11 is not empowered to innovate, it must refer back to current practice.

No European National Standards nor UIC Standards exist to offer a structured process for the design of a wheel. Consequently WG11 referred to the criteria and assessment methods of ERRI Committees B169 and C163. Regrettably there was no UK representation on the B169 committee since BR declined to participate, which precluded participation of the UK supply industry.

The process adopted allows the user to specify the wheel's service environment, after which there is a four step method of approval:-

- Geometric, considering the scope for interchangeability with previously accepted types;
- Thermo-mechanical, to confirm strain levels are acceptable under the designated braking environment;
- Mechanical, to confirm the fatigue strength of the wheel web;
- Acoustic, comprising reference against an existing wheel with known characteristics under the chosen operating environment.

The thermo-mechanical, mechanical and acoustic steps allow progressive assessment using calculation, rig testing or on-track testing. This gives the designer the flexibility of choosing a design which is robust, based on past practice and can be evaluated by the cheapest option (calculation) or to adopt much tighter and more closely optimised design margins which require validation by the more expensive testing options. This draft Standard is anticipated to go to National Enquiry by the end of 1999.

6.1.2 Axle Design Methods

There is no UK National Standard for the design of axles, although there are well established and proven former British Railways Standards that date back to 1968. Within Europe ERRI Committee B136 carried out their own assessment of axle design methods, with BR participation. In 1979 they published a design method which followed similar principles to the already published BR method but contained significant differences in the determination of the influence of environmental effects. This method was incorporated into the National Standards of France, Germany and Italy, where its use, under the conditions prevailing there, has proved to be successful and safe. Consequently, as an existing proven National Standard, it formed the basis for the new CEN draft Standard.

Like the BR method, the axle geometry is calculated to sustain a derived overall fatigue load case for infinite life. The maximum permissible stress, however, is some 40 per cent lower for an equivalent axle material because European practice has been to work with a much lower corrosion allowance. In the light of the Rickerscote investigation, this practice is questionable within the UK operating environment. To safeguard the UK once the Standard comes into effect, clauses have been agreed which require the designer to consider the overall maintenance and corrosion protection regime in choosing his maximum permissible stress.

Similarly, since the Rickerscote investigation questioned the quality of track within the UK, and the past proof of the method has been on European tracks, a clause has been inserted requiring the designer to consider the influence of the infrastructure in selecting the maximum permissible stresses.

The two Axle Design Method Standards are anticipated to be put to National Vote before the end of 1999, they have UK support.

6.2 Wheelset Product Standards

Three product Standards have also been prepared:-

- Wheels, Product Requirement (prEN13262)
- Axles, Product Requirement (prEN13261)
- Wheelsets, Product Requirement (prEN13260)

All three documents follow the format described above. They have passed through the National Enquiry stage and are due for National vote by the end of 1999. In their present format the three draft Standards will incur a negative vote from the UK, because they make third party certification mandatory, which is contrary to BSi policy.

6.2.1 Wheels, Product Requirement

In order to apply the Standard to as wide a market as possible, WG11 decided to limit the initial product Standard to the most commonly used wheel type in Europe. Consequently it defines the requirements for rim chilled, forged and rolled monobloc wheels manufactured from vacuum degassed steel. This type is considered to cover some 80 per cent of the European market.

Four steel grades have been designated : ER6, ER7, ER8, ER9. The prefix 'E' designates Europe, the remainder of the designation closely matches existing BS, UIC and ISO steel grades. There are, however, certain differences, particularly a reduction in the maximum permitted sulfur, phosphorous and hydrogen contents. The reduction in sulfur content has raised concern within the UK and to meet these concerns WG11 has recently agreed to greater precision in the assessment of hydrogen content.

Other significant differences from the equivalent BS5892 : part 3 relate to fatigue assessment of the wheel web, measurement of residual stress at the wheel rim and assessment of fracture toughness at the wheel rim.

Fracture toughness is a controversial issue for the UK. It resulted in a UK negative vote being tabled at National Enquiry because of the concerns of the sole UK wheel manufacturer that the parameter used, a derived value rather than a fundamental property, could be achieved or exceeded by reducing a wheel's ductility and charpy impact resistance. The UK's concerns were supported only by the Czech Republic. The views of the rest of Europe are that the criterion has already been successfully adopted by European railway administrations following a series of reports from ERRI Committee B169, a body in which, as previously noted, the UK declined to participate.

Qualification of the product encompasses two stages, initial supply following laboratory testing, when batch consistency is confirmed, followed by confirmation of satisfactory quantity in service operation.

6.2.2 Axles, Product Requirement
Three standard steel grades, EA1N, EA1T and EA4T, have been designated for axles, where the designations are closely related to existing material used in Europe and the UK. The scope of the Standard applies to solid and hollow axles, made of vacuum degassed forged or rolled steel.

Fatigue test requirements are specified, together with limiting residual stress values and specific corrosion protection requirements. Following a UK request, these requirements permit a waiver of corrosion protection if the maximum stress can be demonstrated to be 60 per cent, or less, of the maximum permitted stress, thus regularising within Europe current UK freight practice for new axle designs.

Qualification is restricted to laboratory tests, the routine delivery tests performed on each axle obviate the need for supply phase qualification.

6.2.3 Wheelsets, Product Requirement
The wheelset Standard applies to a fully assembled wheelset with constituent parts (wheels, axle, brake discs, bearings) complying to the relevant EN Standard. Like BS5892: part 6 it defines the requirements for the finished product but, unlike the BS it defines fatigue characteristics for assemblies.

Qualification encompasses the four steps described above but supply qualification is restricted to a quality assurance basis.

7. FUTURE DEVELOPMENTS

In addition to the draft Standards that are proceeding to National vote or Enquiry during 1999, two more Standards are at working draft level in WG11:

- Wheels - Rim Profile
- Wheels - Maintenance Requirements

7.1 Wheel - Rim Profile
This document is scheduled to go CEN Enquiry during 1999. It has been produced in response to the Operability Directive and defines a standard wheel width, flange shape and tread run out zone. Initially the tread profile was not defined and this was supported by the UK since this was considered to be appropriate for a performance requirement, being dependent on suspension characteristics as well as the chosen railway infrastructure.

However the concensus view was that existing profiles for high speed should be detailed on an Informative, rather than Mandatory, basis. Consequently the UIC 1002, DB 1/40 and BR P8 (now designated EP8 to reflect minor changes to conform with the standard rim) are now cited in the draft.

7.2 Wheels - Maintenance Requirements
Whilst the preceding draft Standards are all concerned with procurement, the significant number of references to the need to take account of maintenance requirements within those

documents has identified the need to define baseline maintenance requirements. The UK has defined a structure based on Railway Group Standards which has been enhanced by specific values contributed by SNCF. CEN/TC 256/SC2 has started the process of enquiry with National Standards bodies of whether they will support the production of this proposed Standard.

8. CONCLUSIONS

The emerging European wheelset Standards differ from the existing British Standards because their scope is significantly greater. The combination of Railway Group Standards, British Standards and industry standards did cover these requirements but use of this fragmented approach would not have permitted equitable procurement within Europe and would eventually disadvantaged home suppliers and purchases. The format of the new Standards may be unfamiliar today but users and suppliers will find them indispensable in the future as the European rail market expands.

ACKNOWLEDGEMENTS

The author would like to thank Mr A.W. Butler, Managing Director of the engineering **link,** for giving permission to publish this paper and Mr J. C. Tourrade of SNCF, Convenor of CEN/TC256/SC2/WG11, for his support in its preparation.

Axle inspection: the use of Monte-Carlo analysis and risk assessment to optimize axle inspection interval

J A BENYON and **A S WATSON**
Traction & Rolling Stock, AEA Technology, Derby, UK

SUMMARY

The depot-based ultrasonic inspection of axles is a time consuming process, which increases the risk of bearing failures because of the necessity of removing and refitting bearing end-caps. If the axle examination process could safely be extended to wheelset overhaul periods, when the axle was stripped, operating costs would be reduced, vehicle availability would be increased and there would be no increased risk of bearing failure because of end cap removal and replacement.

This paper utilises Fracture mechanics, Monte-Carlo simulation and risk assessment methods to define a methodology for setting the NDT inspection periodicity of axles. The objective is to show that there is a technical justification for extending axle inspection, in some cases to wheelset overhaul intervals, while retaining the balance of acceptable risk and maintenance cost. The method described shows that the introduction of Magnetic Particle Inspection (MPI) with its greater sensitivity has significant advantages over Ultrasonic Axle Testing (UAT) methods.

1. INTRODUCTION

A railway axle is not a damage tolerant component. Service failure carries a very high risk of derailment with potentially disastrous results for life and property. The UK railway network experiences typically one or two such failures per year. Analysis of recent axle failure data shows that the majority not directly attributed to bearing problems are caused by fatigue that initiates from corrosion pits or other damage on the axle surface. It is thus vitally important that axles are withdrawn from service before any fatigue cracks can grow large enough to cause the axle to fracture.

To this end, NDT inspection of all axles is carried out at regular intervals. The inspection periodicity can be determined by assuming that a small crack is present and using Linear Elastic Fracture Mechanics to predict the time required for the crack to grow to a dangerous size; this calculation forms the basis from which to define the inspection interval. The better the resolution of the NDT method, the greater the probability of detecting the crack before it can grow to a dangerous size. Thus the NDT inspection method plays a major part in the process. Moreover, a single prediction of inspection interval is of limited value as there is no indication of probability of failure. The use of a statistical process such as a Monte-Carlo method, offers the chance to undertake a quantified risk analysis to establish a basis for inspection periodicity.

1.1 Principle of the Improved Method

The improved method is based on the premise that the principle of ALARP (As Low As Resonably Practical) should be applied, and the "cost per life saved" for the new regime should be equal to or greater than the current Railway Group Safety Plan (Ref. 1).

The approach requires the use of fracture mechanics combined with Monte-Carlo simulation. A large number of crack growth calculations are carried out, covering the likely ranges of the various inputs to the crack growth model, in order to build up the distribution of failure probability as a function of mileage. If the probability of detecting a defect at a given size is also included at this stage, a probability of failure versus inspection period is determined. The information from this calculation is fed into a risk analysis to assess the cost of axle failure and inspection cost.

2. FATIGUE MODEL

The fatigue model utilises standard fracture mechanics to define the crack growth rate (da/dN) in terms of the stress intensity factor range (ΔK in $MNm^{-1.5}$), a geometrical term known as the compliance function (Y), and the crack length or defect size (a) as follows:

$$\Delta K = \Delta\sigma \times Y \times \sqrt{\pi \cdot a} \quad (1)$$

$$\frac{da}{dN} = C \times (\Delta K)^m \quad (2)$$

where C and m are material parameters known as the Paris constant and slope respectively. In the case of cracks growing in axles, the value of the compliance function changes as the crack grows. The crack growth law described by equation (2) applies only where the stress intensity factor is neither so large that the crack growth accelerates toward fracture (not relevant to axles) nor so small that non-linear threshold effects apply. With axles, the loading is often close to threshold and the ΔK is modified to $\Delta K_{effective}$ (Ref. 2)

It is important to know which input parameters are the most sensitive in determining the life of the axle. Studies have been carried out where one fatigue parameter was varied at a time, with all others remaining unchanged. The results of the analysis showed that the threshold and initial crack sizes are critically important, the applied stress almost as significant, whilst the Paris parameters and Fracture toughness were less so. The analysis shows that for an accurate analysis to be carried out the distribution of the most sensitive parameters needed to be quantified.

2.1 Initial and Final Crack Depths

The initial crack size assumed to be present in any axle is that which is just below the detectable size with the NDT method used. For UAT this has historically been taken as 5 mm although there is evidence (Ref. 3) to suggest that a more reliable figure is 10 mm. For MPI a figure of 2 mm is normally used. Both processes also have associated probabilities of failing to detect defects. MPI is considered to be more reliable and has a lower probability of failure to detect a defect. The precise value of the final crack size is not particularly important, as for relatively large cracks, the crack growth is so rapid that failure is imminent whatever the actual crack depth. A 20 mm crack depth is commonly used to define failure for axles.

2.2 Applied Stress

The stress acting on each axle is also one of the variables that is not well defined. Each different design of axle is likely to see a different spectrum of stresses for a given length of track as a result of different axle geometry, axle loading etc. The development of reliable telemetry systems has greatly simplified the measurement of axle stresses and we strongly recommend that the axle environment be defined by carrying out a measurement exercise wherever possible. This should include the usual variations in payload normally experienced. For cases where measurements are not possible, scaling existing strain histories is possible but less accurate and thus less reliable.

2.3 Crack Growth Parameters and Functions

The Paris constants and threshold should ideally be determined from fatigue tests on complete axles, as tests have shown significant differences in the crack growth relationships between small specimens and complete axles. (Ref. 4) Thus, data derived from full size axles are used for all calculations.

No threshold data are available for full size axles. Tests on fully reversed specimens have been carried out by AEA Technology to define mean and standard deviation threshold values.

Fracture toughness data are not available for full size axles. Where necessary, a generally accepted value of 45 $MNm^{-1.5}$ is used, and scatter is not considered.

Observation has shown that an established crack in an axle is usually semi-elliptical in shape. The aspect ratio of the crack (depth / surface length) will vary but it can be approximated by a constant value of 0.8. Raju & Newman (Ref. 5) have developed a compliance function based on finite element work for a cracked bar in bending which is used for most cases.

2.4 Monte-Carlo Simulation

A computer program incorporating a crack growth model has been written to carry out the Monte-Carlo analysis. The input parameters are defined by statistical distributions, and the program performs a very large number of crack growth calculations in order to obtain a distribution of lives. Thus a curve of probability of failure versus life is obtained. By considering different probabilities of detection associated with different NDT inspection techniques (Figure 1) probability of failure against inspection interval for different NDT methods can be obtained. (Figure 2)

Up to now, the calculation has been based on the assumption that all axles contain a crack-like defect. This is not so, and an adjustment to estimate the probability of failure against inspection interval for axles containing defects has to be made. Two methods are possible to do this. A direct method is to use published axle rejection data compared with population size to obtain a probability of rejection, and thus crack initiation rate. A more complex method is to attempt to model the initiating fatigue process. For simplicity we use service axle rejection data to estimate the probability of any axle developing a defect. The probability of a crack initiating is estimated by comparing the axle rejection records for passenger vehicles with the numbers of axles from vehicles in service as follows:

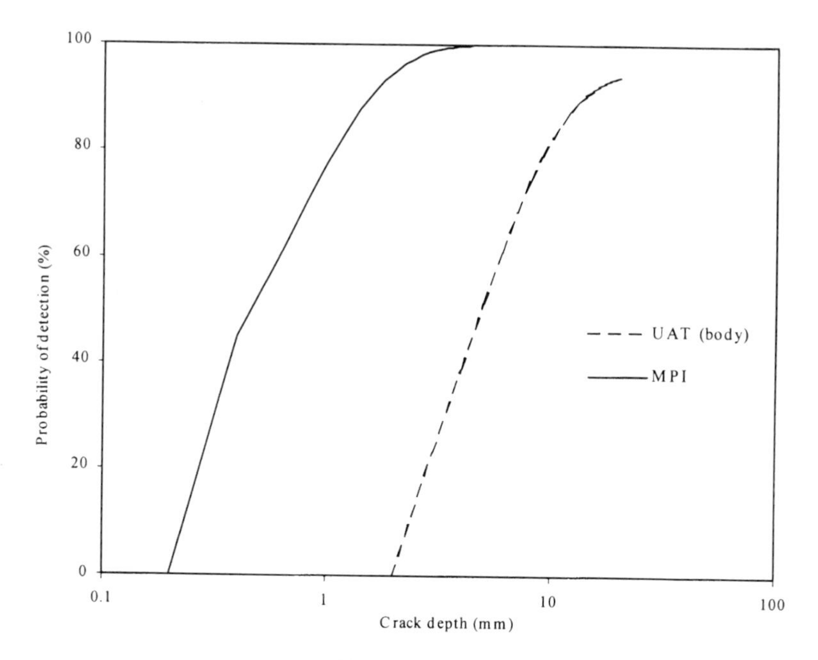

Figure 1

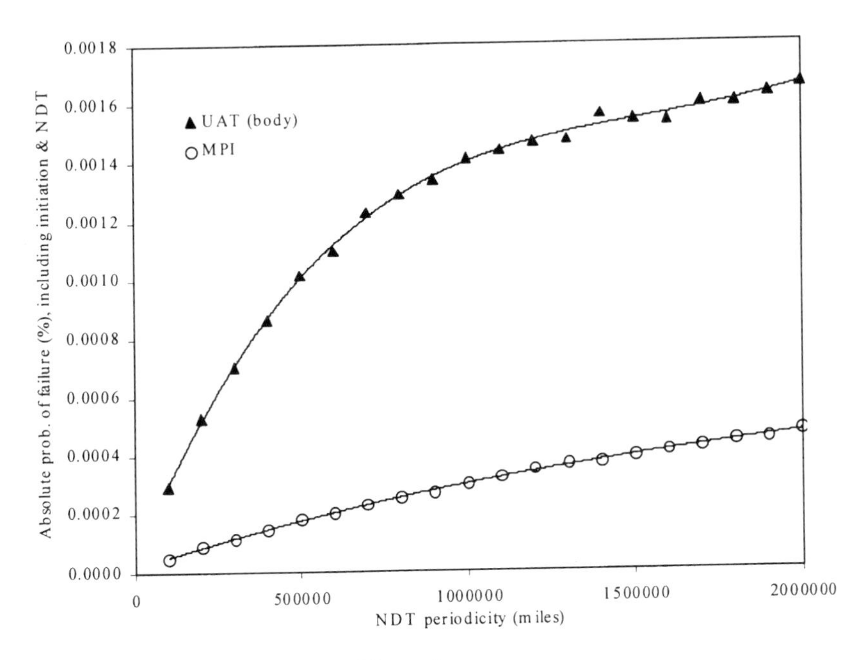

Figure 2

Location	Failures 1980-1995	Confirmed cracks 1980-1995	Average axles 1980-1998	Axle cracks per year	Initiation prob.
Axle body	2	105	65188	6.69	0.000103
Wheelseat	4	8079	65188	505.19	0.00775

The initiation rate for wheelseats is much larger than that for the axle bodies. This is due to the fretting that is commonly found in wheelseats, and which also occurs in gearseats. The initiation probability for gearseats is thus equal to that for wheelseats.

For axles with stresses higher than those allowed by BASS design codes, the probability of initiation is increased; a reasonable estimate of the degree of increase is the overload factor raised to the power of the fatigue S-N slope, which is 6.60 based on fatigue tests of used axles.

3. RISK ASSESSMENT

Whilst the Monte-Carlo Fracture Mechanics approach is much more useful than a single shot analysis, it in itself cannot be used to set inspection periodicity, as no guidance is given as to the acceptable failure rate. Some form of risk analysis is thus required, along the following lines:

(i) Determine the probability of an axle failing using the method defined above.

(ii) Assess the probability of derailment if an axle does fail.

(iii) Determine the cost of derailment, in terms of casualties and damage.

(iv) Determine the cost of inspection.

(v) Combine the above inputs together to determine the optimum inspection periodicity.

3.1 Probability and Cost of Derailment

The probability of derailment following an axle failure is taken as 1.0. There are two costs associated with a derailment. Firstly there is the possibility of death or injury, expressed as "equivalent fatalities", converted to cost using Railtrack figures of £2.65 million per equivalent fatality, and secondly, there is the financial cost of damage to the vehicles and infrastructure plus that of disruption to the train service. The material damage cost is set at £1.3M for each equivalent fatality. Both costs are a function of the vehicle speed.

The number of casualties has been evaluated using a relationship between equivalent fatalities and speed, based on the accident at Polmont, which gives

$$\textit{Worst case casualties} = 0.2727(V - 30)$$

where V is the train speed in mph. It is assumed that no casualties occur in derailments at a speed of less than 30 mph.

3.2 Cost of Inspection

The cost of inspection is made up of two components, i.e. the cost of the labour required to carry out the inspection, and the cost of capital equipment, training etc. which is spread over all the axles inspected. Labour normally represents the majority of the total cost.

An additional cost that could be included is that of "false alarms"; for instance ultrasonic inspection often flags axles up as suspect, but more detailed examination shows that there is no crack present. In such cases the cost of the unnecessary further investigation should be considered when calculating the overall cost of inspection.

3.3 Optimum Inspection Periodicity

The cost of inspection and the cost of derailment are two functions that can be added together to produce a total cost curve. A total cost curve is produced for both NDT methods. (Fig.3 & 4). The minimum costs represent the optimum inspection interval for either NDT method.

4.0 Discussion

The example shown above refers to a modern Electric Multiple Unit. The methodology enables inspection intervals to be set using standard techniques and Railway Group Plan cost values. It also allows different NDT methods to be compared in a quantitative way. By varying key fatigue parameters and costing data sensitivity studies can readily be carried out in order to obtain robust solutions. The model is sensitive to particular key parameters such as axle load, and care should be taken to ensure appropriate data is used. Some simplifying assumptions are made, (such as the estimate of the defect initiation rate), but in general the better the definition of the input data (including if necessary track measurements) the more reliable the answers obtained.

The cost data used to obtain the curves has been based on estimates and will clearly influence any conclusions. Some agreement across the industry on cost consequence models would help to standardise the methodology. However, taking the life of a wheelset (i.e. the wheel wear life) as around 500,000 miles the cost curves suggest in the case above, that the use of MPI methods is advantageous over UAT because axle life costs are lower. In the example shown, the longer the life of the wheelset, the more attractive MPI becomes as the costs continue to fall.

It is interesting to note that the inspection intervals predicted above indicate longer periods that would be specified by MT307 (Ref. 6), suggesting that compliance to the standard may be unduly conservative. This level of conservatism can be costed by subtracting the actual UAT inspection interval cost from the wheelset overhaul period cost.

5.0 CONCLUSIONS

I. Monte-Carlo simulation and risk analysis can be used to optimise inspection periodicities for a given axle and NDT method.

II. The cost of alternative NDT methods with different detection levels can be established objectively.

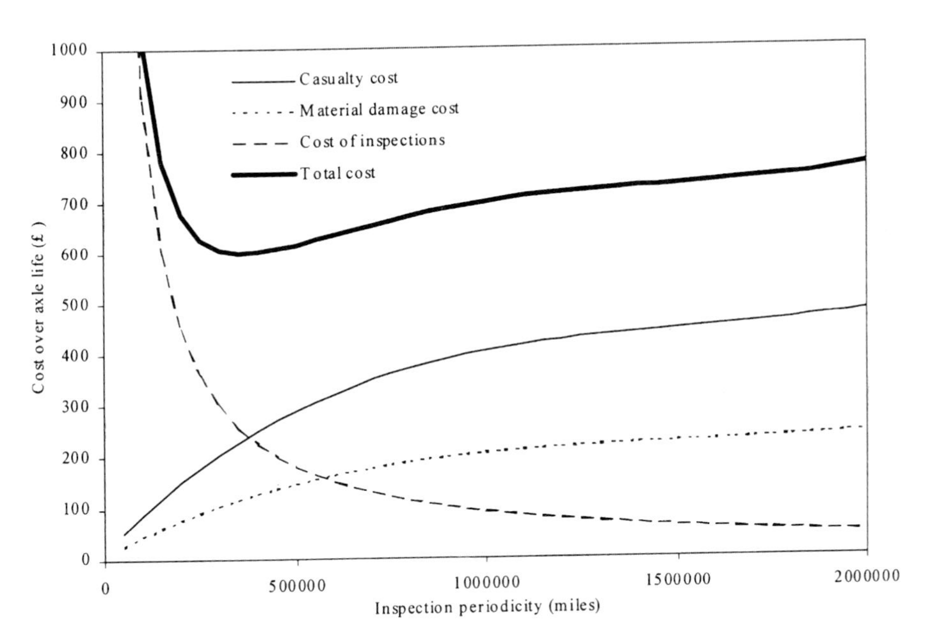

Figure 3 - UAT inspection

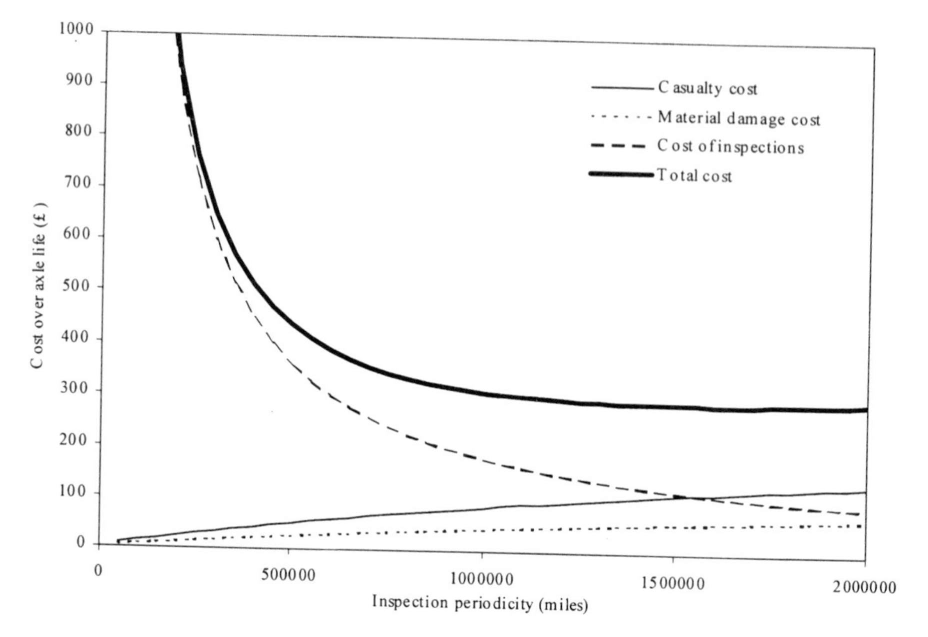

Figure 4 - MPI inspection

6. REFERENCES

1. The Railway Group Safety Plan 1998-99. Railtrack
2. Schutz W, "The Use of Fatigue Thresholds for Engineering Applications", Fatigue Thresholds International Conference , Stockholm June 1981, published by EMAS, pp 911-927
3. B Hawker and A Watson. Axle Integrity Management Review following the HMRI Rickerscote Report: Task 2 Review of NDT Methods. AEA Technology report 2538 February 1998.
4. JA Benyon Factors Affecting the fatigue Life of Axles, BRR Report RR-STR-94-094, December 1994.
5. Raju I S & Newman J C, "Stress Intensity Factors for Circumferential Surface Cracks in Pipes and Rods under Tension and Bending Loads", ASTM STP 905, July 1986, pp 789-805
6. MT307. Frequency of Non-Destructive Axle Testing. Issue 1 Revision A February 1991.

S629/004/99

TreadVIEW™: automated wheel profile measurement

J POCOCK
AEA Technology Rail, Derby, UK

SYNOPSIS

Machine vision systems offer a number of possibilities for automating and improving many aspects of railway inspection and maintenance. One such application developed by AEA Technology Rail is of particular relevance to wheelsets. TreadVIEW™ is a fully developed product that is already being used to automatically measure wheel profiles and make consistent recordings of critical profile characteristics. This paper describes the technical features of this system, shows some typical results, discusses system implementation and benefits, and examines some associated applications.

1. INTRODUCTION

AEA Technology has a number of year's experience in designing, installing and supporting machine vision measurement systems for railway applications. Initial systems concentrated on automating brake pad inspection, but more recent systems have expanded the capability to wheelsets, brake blocks and pick-up shoes. This paper describes the development and implementation of TreadVIEW™, an automated wheel profile measurement system.

2. TREADVIEW™

TreadVIEW™ is a trackside system that automates the inspection of wheel profiles by taking pictures of the illuminated profile as the train passes slowly over the system, and then gauging the resulting image. The results are downloaded to a database that uses the measurement history to determine the wear rate of individual wheel parameters and provide extensive additional fleet information. As well as reducing manual inspection time, TreadVIEW™ allows a more predictive approach to maintenance and provides a comprehensive, consistent and reliable method of making fleet wide profile measurement.

2.1 System Description

The system comprises two elements; a trackside data collection system and an office based information display.

The trackside equipment consists of two pairs of camera and laser combinations mounted beneath the track, and two wheel sensors clamped to the rail. The lasers are used to illuminate the wheel profiles, with the cameras taking pictures of each profile in turn as the passing wheels activate a wheel sensor. The images are collected and analysed by a trackside computer. The system is non-contacting and entirely solid state with the exception of washers and wipers fitted to the camera and laser enclosures to keep the top glass clean.

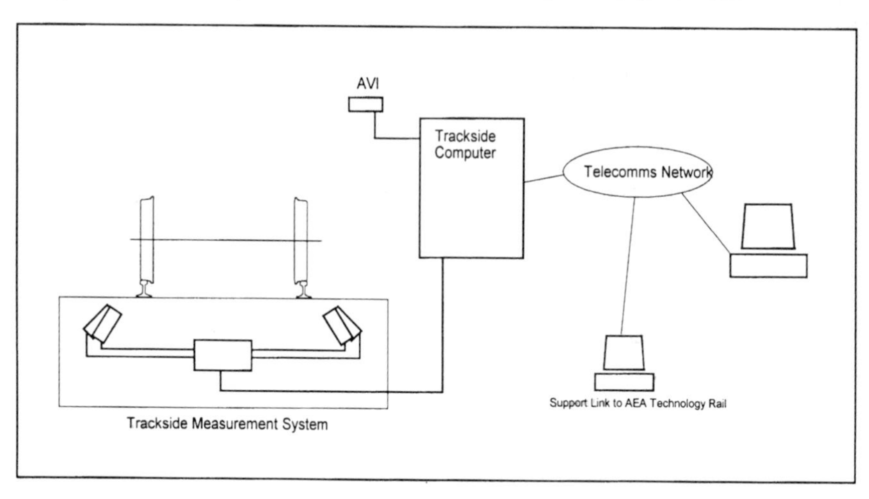

Figure 1 : TreadVIEW™ system schematic

2.2 Installation Requirements

The trackside measurement system is typically mounted in two 500mm deep troughs, which are installed between two sleepers in ballasted track. The only modification required to the track form is the extension of the existing sleeper spacing to 750 mm in order to accommodate the troughs. The troughs are designed to take ballast loading and be self-draining. Installation does not affect track alignment or drainage and no modification to the rails is required.

The trackside computer is typically housed in a weather proof industrial rack which can either be installed at the trackside or in a nearby plant room if one is available.

No modifications to the rolling stock are required to make the wheel profile measurements. Vehicle identification is usually carried out using an additional camera, provided that a consistent, legible and unique car identification mark or number is visible.

The only routine maintenance required is to replace screen wash fluid and to ensure the equipment is kept free of debris and excessive dirt.

2.3 Data Collection

The trackside system is designed to operate automatically. As a train approaches the measurement equipment the leading wheelset triggers a wheel sensor which initialises the system. The camera and laser enclosures are cleaned using the wash wipe system, and the power to the lasers is switched on. As the train passes over the cameras and lasers, further wheel sensors are used to trigger the laser beam illumination and image capture for each wheel profile. The images are collected and saved for analysis after the train has passed.

Each illuminated wheel profile is created by a single laser line projected radially onto the wheel at a nominal angle of 45 degrees. The camera views the reflected light from a position perpendicular to the rail though inclined towards it at a nominal angle of 45 degrees. The camera is fitted with a band-pass filter to remove all light except that at the laser frequency giving good rejection of all but the strongest ambient lighting.

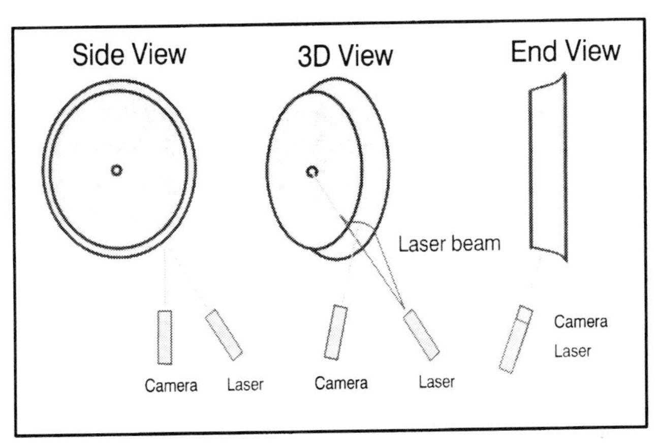

Figure 2 : Camera and laser arrangement

2.4 Image Analysis

The resulting image consists of a white stripe against a black background (shown on the left-hand side of Figure 3). This image gives only a general impression of the wheel profile because of skewing caused by camera and laser angles. To correct for this, co-ordinates of the line profile are extracted from the image and then subjected to a three dimensional co-ordinate transformation before being processed more fully to extract specific wheel profile parameters.

The transformed profile and wheel profile parameters for each image are saved to a result file, which is downloaded automatically by the Information Display System.

2.5 Automatic Vehicle Identification

Automatic Vehicle Identification (AVI) is required in order to track individual wheel profiles over time. Both current TreadVIEW™ installations use an additional camera to take pictures of the side of the passing vehicles. These images are analysed to extract the vehicle and unit numbers, and this method has proved to be a reliable and robust way of providing AVI at marginal cost without the need to fit tags to vehicles. Figure 4 shows a Northern Line vehicle number 51581 being identified at the VIEW™ installation at Morden Depot.

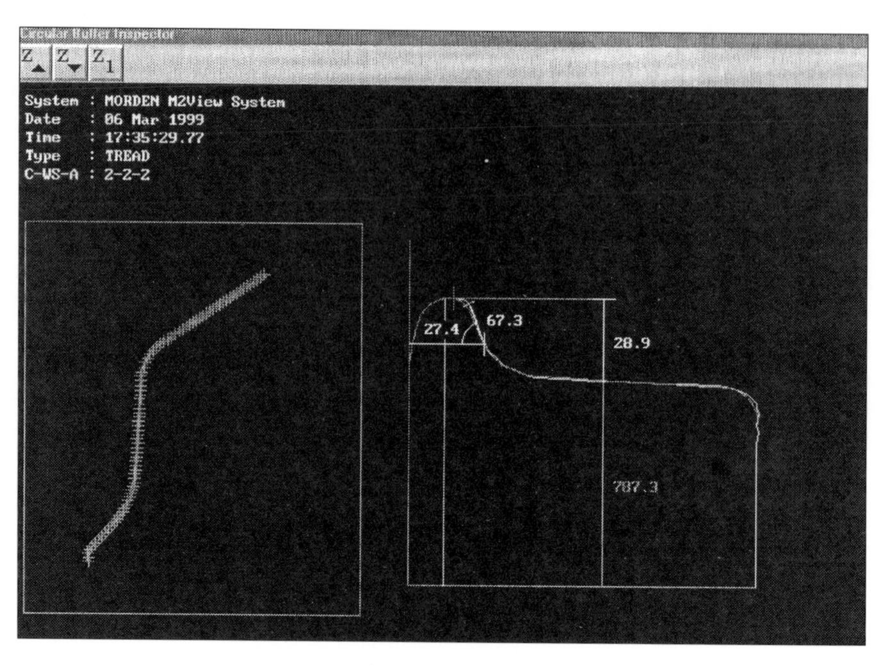

Figure 3 : TreadVIEW™ image analysis

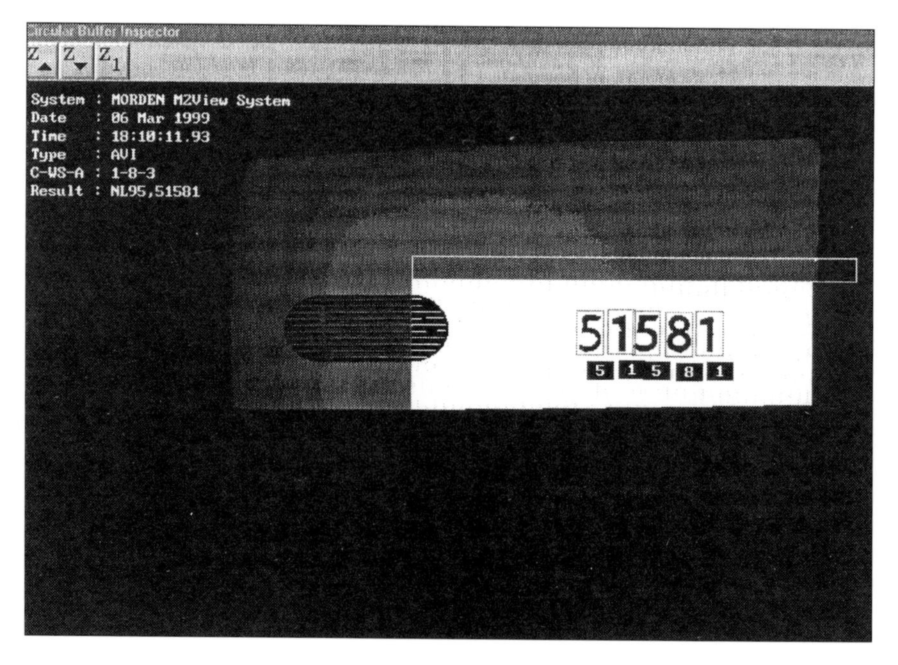

Figure 4 : AVI image analysis

3. OFFICE DATABASE

The results from the trackside system are automatically downloaded each day to an information display system. The display system provides all the individual measurements to the user, along with long term histories. The wear rates of individual wheelset parameters are calculated, allowing predictions about the remaining wheelset life to be made. Additionally, the user can search for particular ranges of parameters at individual axle positions and across cars and units, as well as viewing the state of the whole fleet at once.

The office system allows maintenance staff to monitor and manage wheelset maintenance over the entire component life.

3.1 Maintenance Printouts

Individual vehicle printouts can be generated to summarise the state of the wheelsets on a particular vehicle. These sheet are intended to be used when a vehicle comes in for maintenance to highlight what needs to be done and to identify any additional inspection work that is necessary.

3.2 Problem Identification

Database searches can be carried out to identify wheelsets with parameters that are approaching or exceeding limits. This allows individual wheelset problems to be highlighted such as thin flanges or hollow wear. Figure 5 shows a wheel profile with severe hollow wear identified using a search for wheelsets with very high flange heights. The upper profile shown is the idealised profile.

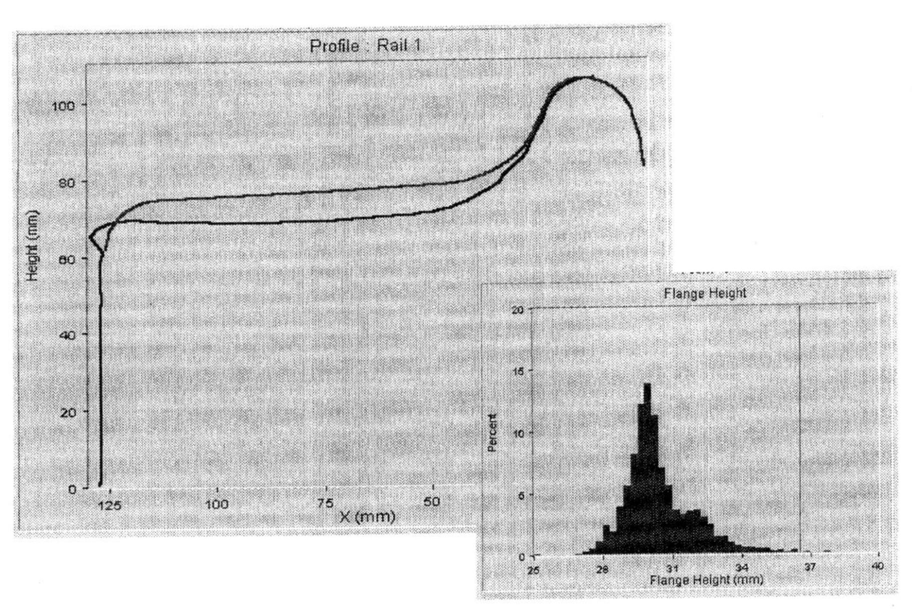

Figure 5 : Severe hollow wear identified by searching for wheelsets with large flange heights

3.3 Profiles and Wear History Plots

Plots showing the wheel profiles and parameter wear trends can be generated for each wheelset. This information is useful in allowing some additional human input into the analysis of the results and gives the user greater confidence in the information provided. Figure 6 shows a comparison between a typical TreadVIEW™ measured wheel profile and a MiniPROF plot, along with a wear rate history of the wheelset flange height over a three-month period.

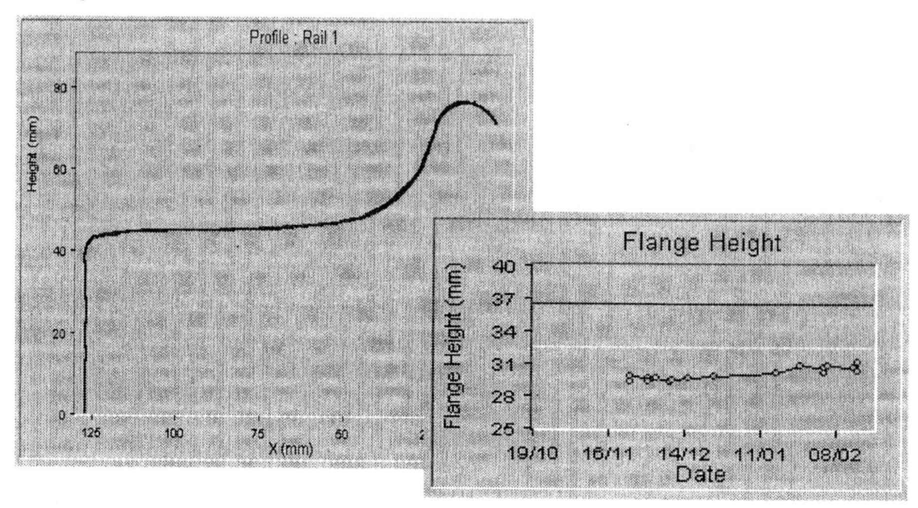

Figure 6 : Profile and wear rate history plots

3.4 Statistical Reports

Statistical reports showing the distribution of parameter values and wear-rates by axle, vehicle, unit and across the whole fleet can also be generated. These are particularly useful in providing a more global view of the fleet performance and a longer-term assessment of the fleet maintenance requirements. Perhaps the most powerful feature is the ability to combine these distributions with knowledge of the individual parameter wear rates to project the distribution forward in time. In this way a prediction can be made about the number of wheelsets that will exceed maintenance limits in the next year. Such information could be invaluable in terms of determining the expected maintenance budget, assessing materials usage, and identifying possible large-scale fleet problems before they affect the service.

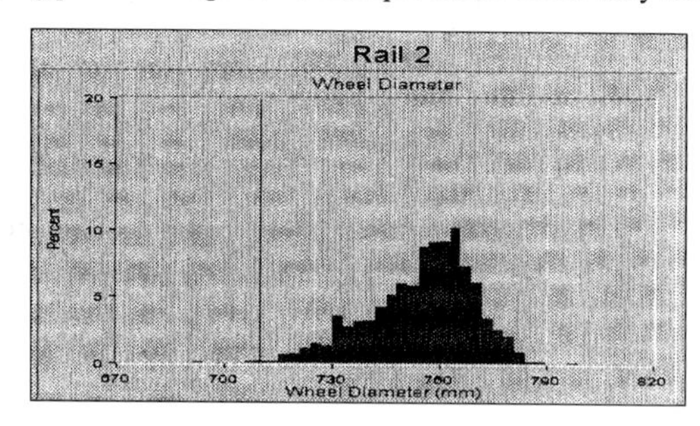

Figure 7 : Typical wheel diameter fleet distribution.

4. TREADVIEW™ IMPLEMENTATION AND RESULTS

The first TreadVIEW™ system was installed at the London Underground Victoria Line depot at Northumberland Park Depot in September 1997. The system is installed on the approach to the train wash and sees around 40 train passes a day.

The first production system was installed as part of a much larger VIEW™ installation at the Alstom Northern Line depot at Morden in February 1999.

4.1 Pre-production Experience

The Northumberland Park installation was intended as a pre-production demonstration, since there were a number of technical issues associated with making measurements outside from a passing vehicle, which it was not possible to assess in the laboratory.

The main concerns were that the system would not be able to cope with excessive levels of ambient light and that variations caused by vehicle movement and environmental conditions would reduce the system accuracy.

As expected, in practice it has been found that the most common problem for the pre-production system is stray reflections caused by bright and low sunlight conditions, and poor reflection of the laser line on the tread surface. These issues have been dealt with at Northumberland Park by decreasing the camera shutter speed to see more laser light, with a slight trade off in system accuracy caused by the very minor image blurring. This problem has been overcome in the production system by using lasers with greatly increased power output, which allows a much high shutter speed to be used without losing any laser line definition.

Although there are some slight losses in accuracy caused by taking measurements of a wheelset on a moving vehicle, these have not proved to be significant due to the regular measurements made. Experience has shown that whilst it is not always possible to analyse all images, this does not matter as there is a high probability of the wheelset being measured the next time the train passes over the system. Since wheelsets tend to wear over a matter of months rather than weeks, the system is very robust its ability to cope with such eventualities. On average, a wheelset in measured at least once every 4 or 5 days.

4.2 Northumberland Park Results

TreadVIEW™ has been in use by Northumberland Park depot since January 1998 as the primary method of wheel profile measurement. Over the three-month period prior to this the entire Victoria Line fleet had been measured once using a portable profile-measuring device called a MiniPROF. This provided a very accurate baseline for comparison with the TreadVIEW™ results.

Results from this exercise have demonstrated that it is capable of making individual measurements at speeds of up to 12 mph with the following accuracy:

Flange Height	±1.0 mm
Flange Width	±1.5 mm
Wheel Diameter	±4.0 mm

Individual measurement repeatability from a stationary wheelset is of the order of ±0.3 mm.

One of the great strengths of the condition monitoring approach used with TreadVIEW™ is that the measurement of each wheelset occurs much more frequently. The measurement accuracy can therefore be improved by averaging results over time. Statistically it can be shown that the measurement error reduces by the square root of the number of measurements taken. For example, after 9 passes over the system, the error would be reduced to one third by averaging all results. In practice, the presence of wear trends and other factors complicate things but a significant improvement is usually obtained. Combined with the additional effect associated with increasing the production system laser power the following long-term accuracies are achievable:

Flange Height	±0.3 mm
Flange Width	±0.8 mm
Wheel Diameter	±1.5 mm

Flange Width measurement accuracy is worse than Flange Height because the method of measurement is affected by the wheel width, which carries a manufacturing tolerance. The accuracy could improve by using two lasers and cameras per profile. It is estimated that this method could achieve long term Flange Width accuracies of ±0.3 mm if required.

Wheel Diameter measurement accuracy is hampered on the Victoria Line due to the lack of a visible turning groove, but could be expected to be within ±1.0 mm if a groove existed.

5. ASSOCIATED APPLICATIONS

5.1 Combined Automated Vehicle Inspection

Wheel profiles are not the only under-vehicle component that require regular inspection, and there is great advantage to be made in combining the inspection of more than one vehicle component with a single installation. An example of this can be seen in the latest VIEW™ installation on the Northern Line. This system combines TreadVIEW™ with ShoeVIEW™ (for inspection of pick-up shoes) and BlockVIEW™ (inspection of brake blocks). Not only does such an approach provide additional cost benefit at marginal cost, but with the introduction of new, low maintenance fleets it offers the possibility of carrying out all the regular inspection work automatically and greatly extending the time a vehicle spends away from an inspection shed.

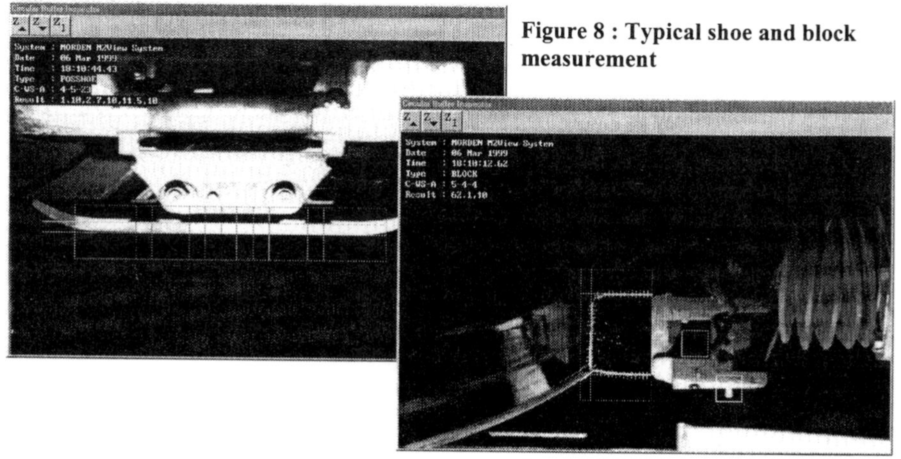

Figure 8 : Typical shoe and block measurement

5.2 Wheelflat Detection

TreadVIEW™ is not capable of identifying wheel flats, since it is highly unlikely that the wheel profile would be illuminated at exactly the correct point. A much more reliable way of detecting the presence of flats and other circumferencial defects is to use a wheel impact load detector such as WheelCHEX™. Such systems use instrumented rail sections to measure the strain induced by a passing axle. By recording the static and dynamic loads induced over a series of sleeper bays it is possible to build up a picture of the entire wheel circumference and spot problems such as wheel flats and out-of-roundness, as well as weighing the vehicle and identifying poor load distribution.

Figure 9 shows a typical loading plot for a vehicle with wheel flats. The bars show the average loading for each axle with the dots showing the maximum dynamic variation. A wheel flat can be spotted when there is a large difference between the two values, such as for axles 9, 10, 11 and 12.

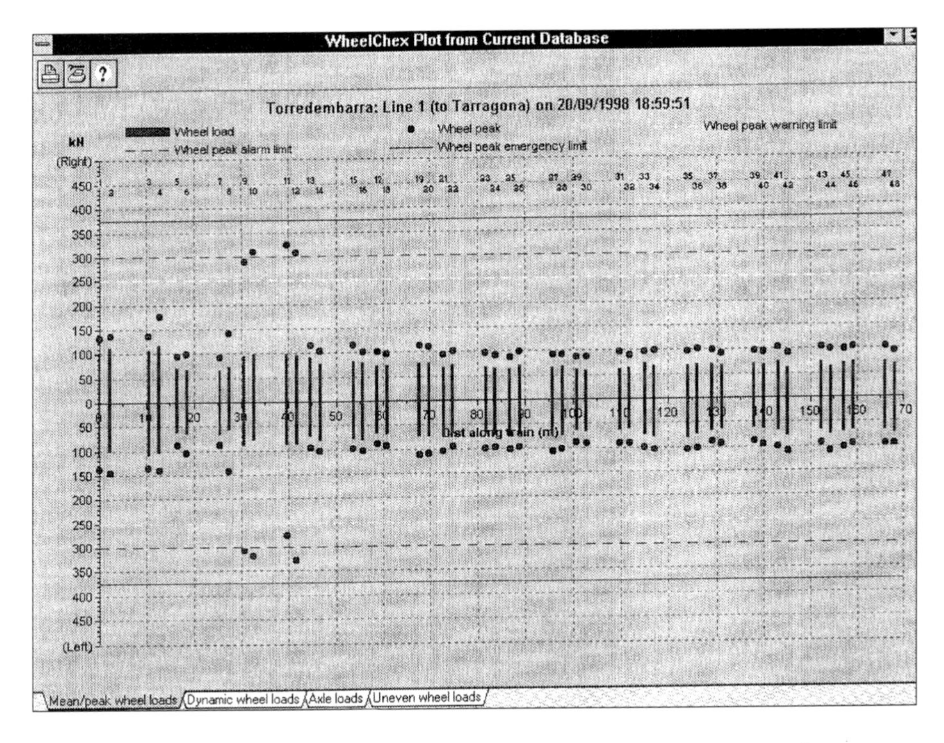

Figure 9 : WheelCHEX™ dynamic and static loads showing wheel flats

6. BENEFITS

The benefits of automatic wheel profile measurement come in five areas outlined below.

6.1 Reduction in the need for Manual Inspection

There are mandatory standards for the post re-profiling and in-service inspection of wheelsets. In service, a number of inspection tasks are required that can be done automatically and more frequently by TreadVIEW™. The manual MiniPROF measurement of the 86 Victoria Line units took about three man-months to accomplish. TreadVIEW™ provided the same information within seven days of going on-line and continues to do so week-in and week-out.

6.2 Identification of Sudden Wear due to Track/Lubrication Faults

Some fleets have few problems with wheel wear, wheels being run until the next bogie overhaul. However, faults with track or flange lubrication equipment can result in rapid wheel wear not always picked up quickly by the depot. Wheelset 'usage' is a high consumable cost and track faults can accelerate wear by a factor of five or more. The early identification and rectification of such faults using TreadVIEW™ can minimise the amount of worn material with consequent savings in wheelset material and the labour involved in inspection, re-turning and unscheduled replacement of wheelsets.

6.3 Better Management of Unavoidable Wear

Some fleets have general wheel wear problems that they have to live with. Unless there is statistically reliable and recent information it is very difficult to manage such a situation, and the problem can easily become unstable if changes to the equilibrium are not spotted quickly. This can be especially true in situations where new vehicles are being introduced onto existing worn track and the distribution of worn wheel profile shapes is suddenly altered.

6.4 Better Maintenance Scheduling and Materials Planning

Depot inspection of wheelsets is often on a pass/fail basis resulting in unexpected requirements for tyre-turning which adversely affects train availability and wheel lathe workload scheduling. By frequent monitoring of wheelset condition and wear rate a predictive maintenance management approach can be adopted, resulting in improved train set availability and better maintenance resource management.

The cost of wheelset maintenance is usually considered to be the largest single cost of all vehicle subsystems. In large part this is as a result of the cost of periodic overhauls, but the cost of routine and unscheduled wheel turning is also significant, due to the amount of material removed from the wheelset (resulting in reduced life) and from vehicle unavailability. A wheelset can typically be turned 3 or 4 times before it gets below its scrapping limit, and on average wheels may be returned every 18 months. This gives a wheelset 'life' of about 5 years and can generate total annual wheelset material 'usage' costs in excess of £2 million for a typical fleet.

The extent or depth of re-turning depends upon knowledge of the profile at the time the turning is carried out. Knowing this in detail in advance would allow more optimal returning to be carried out. This could allow the number of re-turnings to be increased from 3 or 4 to say 4 or 5 in the life of a wheelset. Improving the usage of the asset by only 10% represents savings of hundreds of thousands of pounds per annum.

S629/005/99

Application of North-American railway wheels in United Kingdom freight service

R E BAILEY and **C P LONSDALE**
Standard Steel (USA), Pennsylvania, USA

ABSTRACT

This paper describes the specifications, design, and testing of wrought railway wheels produced for a new type of coil steel freight car and an intermodal container car, both designed for service in the United Kingdom. Relevant specification differences between North American and British wheels are reviewed and discussed. Finite element stress and temperature analyses performed on wheel designs considered for the two car types are compared and contrasted. The paper also discusses some of the important factors in the railway service environment that affect wheel performance. Special attention is given to the thermal stresses which are imposed upon wheels from tread braking.

INTRODUCTION

The privatization of the rail freight business in the United Kingdom and the resurgence of railway freight traffic has resulted in the need for significant additions to railroad infrastructure, including rolling stock (Ref. 1). Due to the need for a new type of freight car to transport coil steel, the English, Welsh and Scottish Railway (EWS) contracted with Thrall Europa to build the cars at York, England. The cars were a new design for United Kingdom service and more closely resemble North American freight cars built to Association of American Railroads (AAR) specifications. This paper describes efforts by Standard Steel to provide forged wheels for the new coil steel car, and a container car that was designed later. Standard Steel, as the supplier of wheels and axles, was part of the bogie redesign and retrofit team for the new car.

The mission of the bogie redesign and retrofit team was to 1) meet schedule dates, 2) satisfy customer expectations (including weight), 3) meet all relevant Railway Group Standards, 4) pass all acceptance tests, 5) provide all necessary design data and materials for the retrofit program, and 6) to provide a product free of field failures. Important wheel concerns included

1) stress levels produced in the wheel by mechanical and thermal loading, 2) temperatures experienced by the wheel due to braking, 3) the amount of wheel lateral deflection produced during brake heating, 4) wear resistance, and 5) spalling resistance. The decision was made to begin the analysis by using a wheel design that is currently produced in North America. The basis for this decision was the many years of successful North American railroad experience using such designs in heavy haul freight service, along with cost concerns.

SERVICE ENVIRONMENT

There are significant differences between the operating philosophies and practices of freight railroads in the United Kingdom and North America. Although a detailed discussion of such practices is beyond the scope of this paper, a brief review of those items that most affect wheels is provided.

North American axle loads are considerably heavier than in the UK, with 286,000 pound (1,272 kN) gross rail load (GRL) quite commonplace, particularly in coal, grain and steel service. This equates to an axle load of 71,500 pounds (318 kN), while the maximum axle load permitted in the UK is 56,228 pounds (250 kN) (Ref. 2). Railtrak has provided a formula that limits the allowable contact stress by the wheel size as follows (Ref. 2):

$$Q = 0.13D, \text{ where } D = \text{wheel diameter in mm, and } Q = \text{wheel load in kN}$$

Although EWS negotiated with Railtrak to increase the 0.13 value to 0.146 for its cars (Ref. 2), this formula necessitates measurement of wheel diameters in service to insure that maximum contact stresses are not exceeded. For example, for a 56,228 pound (250 kN) axle load, the condemnable wheel diameter is 33.75 in. (857 mm). There is no such contact stress restriction in North America and the analogous wheel condemning limits are based upon the rim thickness of the wheel. Condemning rim thickness (thin rim) for 30 and 33 in. (762 mm and 838 mm) diameter wheels is 3/4 in. (19 mm) or less while for 28, 36 and 38 in. (711 mm, 914 mm and 965 mm) diameter wheels is 7/8 in. (22 mm) or less (Ref. 3). Railroad car inspectors carry a small gauge with them in order to perform rim thickness measurements in the field. Wheels can of course be removed for a host of other defects in North America including high flange, thin flange, etc.

Freight cars in North America are exclusively tread braked and composition brake shoes are most widely used, although cast iron brake shoes are still used on some cars. Conversely, disc brake systems have been preferred in the UK, although there also are a significant number of tread braked cars with cast iron shoes, and some with composition shoes (Ref. 2). Tread braking and composition shoes provide a much harsher service environment with respect to heat generation for the wheel, and excessive braking can lead to reversal of the as-manufactured residual compressive hoop stresses in the wheel rim. Air brake systems in the UK have a graduated release feature in addition to the graduated application feature (Ref. 2), whereas current North American air brake systems have a graduated application feature but are released directly.

Other important issues with regard to braking include the use of locomotive dynamic brakes and the brake ratio of cars in service. In North America it is often standard operating practice to use dynamic brake systems, which use the retarding resistance of locomotive traction motors and dissipate the heat generated through radiator grids, as much as possible in order to avoid excessive tread braking. Dynamic braking of locomotives in the UK only provides for the locomotive weight and thus does not make a significant contribution to train braking (Ref. 2). The braking ratio of cars, particularly when cars are empty, also has an important effect on wheels. If the empty brake ratio is too high or adhesion between the wheel and rail is low, wheel sliding will occur. This then leads to the localized formation of martensite patches and subsequent tread spalling. In North America, freight cars are designed to conform to Association of American Railroads (AAR) brake design requirements (Ref. 4). These requirements specify acceptable loaded brake percentage ranges and maximum empty brake percentages for new equipment. However, UK braking performance is determined by the stopping distance and freight equipment must be stopped within 1,040 yards (951 meters) from 60 mph (96.5 km/hour) (Ref. 2).

The implications of differences in the service environment are quite important to the life of wheels and other freight car components. UK wheel removal data were not available to the authors, but North American wheel removal Car Repair Billing (CRB) data are maintained by the AAR. 1997 AAR CRB data showed that 16% (89,981 wheels) of wheel removals occurred due to "shelling/spalling" (why made code 75), 5% (28,639 wheels) occurred due to "tread slid flat" (why made code 78), 2% (10,818 wheels) occurred due to "built up tread" (why made code 76), and 2% (10,375 wheels) were removed due to "thermal cracks" (why made code 74). The number of wheels removed in 1997 for shelling/spalling was 73 times greater than the number of wheels removed for the same defect in 1972. Other major wheel removal causes are wear related and due to associated repairs. For example, in 1997, 37% (203,513 wheels) of wheels changed out were "removed good condition" (why made code 11), 17% (95,155 wheels) were removed because the mate wheel was scrapped (why made code 90) and 10% (55,649) were removed due to high flange (why made code 64) (Ref. 5).

WHEELS, AXLES, AND ROLLER BEARINGS

The components intended for use in the United Kingdom bogie design project are briefly discussed as follows:

- AAR J-36 Wheel (914 mm diameter)
- AAR Classification F Raised Wheel Seat Roller Bearing Axle
- 6-1/2 x 12 W/HDL (165 x 305 mm) Roller Bearing

Wheels are produced in accordance with the AAR M107 standard for carbon steel wrought wheels. The wheel design used for the coil car is the J-36 design, a 36 in. (914 mm) diameter, two-wear wheel having a minimum 2 in. (50.8 mm) rim thickness, gaged at the taping line. The wheels are produced to Class B chemical requirements (0.57%-0.67% carbon) at the request of the customer. In the United Kingdom, British Standard 5892, R8 material is normally used in similar service (0.56% maximum carbon). However, in North America Class C steel (0.67%-0.77% carbon) is almost always used for freight service due to increased wear resistance. The tread profile for the wheels supplied was to the P-5 design, which is commonly used in the UK. The P-5 contour has a double slope tread with a 1 in 20 slope from the flange radius to a point 1.782 in. (45 mm) from the front face of the rim, and then a 1 in 10 slope from that point to the front face of the rim. The heat treatment is rim chilled to a hardness of 277-341 HBN. The J-36 wheel is produced with a low stress, "S" shape plate design.

Axles used were the standard AAR Class F design, 6-1/2 x 12 in. (165 mm x 305 mm) journal size, used for 100 ton (90.7 metric ton) freight service in North America. The axle forgings are heat treated by double normalizing and tempering and have a minimum yield strength of 50 ksi (345 N/mm^2). Axle stresses were calculated using the British Railways Board Bass 504 method (Ref. 6). The actual stresses calculated for the axle load, along with fatigue analysis, validated use of the design for the intended service. In North America, axle failures rarely occur. This no doubt is the result of standardized designs, ample load carrying capability and wheel shop practices established for maintenance and reuse.

Wheels are pressed onto the axle wheel seats using an interference fit of approximately 0.001 inches-per-inch of wheel seat diameter. AAR 6-1/2 x 12 W/HDL roller bearings are applied also by press fitting. The only significant difference in the assembly for use in the UK is that the wheel back-to-back dimension is 53-35/64, +0.079, -0 in. (1,360, +2, -0 mm) in lieu of the 53 to 53.094 in. (1,346.2 to 1,348.6 mm) dimension. This difference of approximately 1/2 in. (12.7 mm) is to accommodate track gage differences.

WHEEL STANDARDS, DESIGNS & REQUIREMENTS

Wheels for EWS applications have typically been manufactured to British Standard 5892, Part 3, using R8T steel, which is a material having similar strength, hardness, and toughness as the AAR M107 Class B material. Comparisons of chemical analyses and mechanical properties are shown in Table 1. Virtually all freight cars operate in North America with Class C (0.67%-0.77% carbon) wheels. Class C material has higher strength and hardness, but has lower ductility and fracture toughness.

Straight plate (high stress) wheel designs similar to those used in the UK have been prohibited in interchange freight service in North America because overheating of the rim, owing to on-tread braking, has been known to cause stress reversal in the rim. Low stress (S-plate) designs have been developed to minimize the stress reversal effects of high thermal and mechanical loading. Figures 2 and 4 show a comparison of the UK and North American plate shapes. In North America, wheel designs must be verified using the AAR S-660 standard which details finite element analysis (FEA) procedures for AAR design approval stress analysis (Ref. 7). This method of analysis, modified by using EWS mechanical loading and thermal loading criteria, was used to determine the suitability of wheels under specific service conditions expected by EWS, such as the Perth-Inverness grade. Before we begin discussion of design verification for several service cases, we briefly mention several other differences in manufacture of the AAR M107 wheel compared to the British Standard 5892 wheel.

- AAR M107 requires shot peening of the plate surfaces to develop beneficial compressive stresses.
- AAR M107 requires ultrasonic testing of the wheel rim and magnetic particle testing of the plate.
- Mechanical testing is required for the BS 5892 wheel, whereas only brinell hardness testing is specified for the AAR M107 wheel.

FINITE ELEMENT STRESS ANALYSES

The procedure used for the analytic evaluation of wheel designs is the AAR S-660 standard. The AAR S-660 procedure was modified for this analysis using thermal and mechanical loads representing conditions for the Perth-Inverness grade. Only elastic stresses were determined, and these values were used to compare potential new wheel designs to those wheel designs known to have satisfactory UK field performance. Specifically, the work determined finite element thermal and thermal/mechanical elastic stress analysis results for a J-36 S plate wheel and a 952.5 mm British Railway wheel, and the results were then compared.

The analysis was performed by Thomas Rusin of Rusin Consulting Corporation using the finite element computer program COSMOS/M (Ref. 8, 9, 10). Figure 1 depicts a cross-section of a wheel, the coordinate system, points of loading and planes of reference used in the finite element model. The models used in our study were constructed from drawings of the wheels. Both new wheel and worn wheel configurations were used and are shown in Figures 2 through 5.

The loading pattern used in the analyses is described below and is displayed in Figures 6, 7 and 8.

$V_1 + L_1 =$ Combination of 56,202 pounds (250 kN) normal and 29,225 pounds (130 kN) lateral load on the front of the flange and at an axial distance of 1.198 in. (30.4 mm) from the back face. See Figure 6.

$V_1 + L_2 =$ Combination of the 56,202 pounds (250 kN) normal and 22,480 pounds (100 kN) lateral load on the front of the flange and at an axial distance of 1.198 in. (30.4 mm) from the back face. Figure 7 shows the location of the $V_2 + L_2$ loads.

$V_2 =$ 160,000 pounds (716.8 kN) normal at an axial distance of 2.7559 in. (70 mm) from the back face. Figure 8 shows the location of the $V_1 + L_2$ loads. The V_2 load represents the dynamic effects of wheel flats.

Interference = Interference fit of 1/1000 inches-per-inch of diameter at the wheel/axle interface.

Thermal = Thermal load is a direct horsepower input into the wheel, applied evenly on the tread of the wheel over a distance of 3-3/8 in. (86 mm) for the J-36 wheel, and 3.1496 in. for the British wheel. The horsepower was determined by a dynamometer test which simulated the Perth-Inverness grade (Ref. 11).

RESULTS AND DISCUSSION

J-36 and British Wheel Designs

The finite element stress analysis (FEA) previously described provides the ability to compare maximum Von Mises stress levels, maximum rim temperatures, and rim thermal lateral deflections for wheel designs used in North America to those wheel designs having proven service in the United Kingdom. Table 2 shows summary data representing three different braking condition cases for the J-36 wheel on the Perth-Inverness grade. Table 3 compares a

J-36 wheel design to a British 952.5 mm wheel design for a 60 mph (96.6 km/hour) drag brake case on the Perth-Inverness grade.

Figure 9 shows the excellent correlation between experimentally determined thermal results and results obtained using finite element analysis techniques. Computer analysis provides for comparison of temperature distribution in the wheels as shown in Figures 10 and 11. Typical results of analysis for Von Mises stress are shown in Figures 12 and 13 for the new J-36 wheel and new British wheel, respectively. The V_l + L_l + Thermal + Interference loads resulted in the maximum effective stress for the various load cases.

Thermal loads greatly contribute to stresses in wheel plate fillet areas. The curved plate (low stress) J-36 design showed lower stress levels than the British design with the loading conditions used in Table 3. Maximum temperatures attained in rim areas resulting from the thermal loading were less for the British design for both new and worn rim conditions. This is a result of the greater mass in the rim of the British wheel, which is approximately 3/4 in. (19 mm) thicker than the J-36 wheel. Thermal lateral deflections of the rim, which reduce the wheel back-to-back dimensions of the wheel set, were essentially the same for the new wheel profiles, but were 1/32 in. (0.8 mm) less for the J-36 wheel worn rim profile.

The maximum and minimum radial stresses were calculated using vertical, lateral, and interference loads. Then, mean and alternating stresses were calculated for each node of the plate surface in the wheel FEA model and plotted on a modified Goodman diagram for both new and worn wheel configurations. Results show that the Standard Steel J-36 wheel has a greater margin of safety below the fatigue limit than does the British wheel in both the new rim and worn rim conditions.

Analyses were also performed to ensure that the combination of thermal and mechanical loads will not cause the wheels to move on the axle when an interference fit of 0.001 inches-per-inch of wheel seat diameter is used. The forces required to move the wheel on the axle were far greater than those of the intended service and are not a concern.

P-33 Wheel Design

Studies were also conducted to determine the suitability of a 33 in. diameter (838 mm) wheel with a 2-1/2 in. (63.5 mm) thick rim for use in 102 metric ton service (Ref. 12). Analytical procedures employed were similar to those used for the J-36 and British 952.5 mm wheels. Vertical and lateral loads were also the same as used for the J-36 and British wheels, namely 56,202 pounds (250 kN) and 29,225 pounds (130 kN), respectively. Thermal loading consisted of a drag brake case from 60 mph (Thermal Case 1) and a stop brake case from 75 mph (Thermal Case 2), both on the Perth-Inverness grade.

The results for the P-33 wheel analysis for Thermal Case 1 indicate high plate stresses and rim temperatures, especially for the worn rim configurations of 3/4 in. (19 mm) rim thickness. Concern about the levels of temperature and stress determined for these loads prompted calculations for greater rim wear limits of 1-1/2 in. (38.1 mm) and 1-3/4 in. (44.45 mm) in lieu of the 3/4 in. limit normally used in North America to condemn 33 in. diameter wheels. Table 4 shows the temperatures and stresses attained for the new wheel profile and for the 3/4 in., 1-1/2 in., and 1-3/4 in. rim wear limit profiles. In North America, 33 in. diameter wheels are normally used in 75 ton (68 metric ton) service.

For the following mechanical and thermal stress evaluations, the worn limit for the P-33 wheel was established at 1-1/2 in. rim thickness, not the 3/4 in. specified in the AAR Field Manual (Ref. 3). A mechanical stress comparison of the P-33 and British 952.5 mm (37.5 in.) wheel designs shows that wheel plate stresses are 18% and 28% greater for the P-33 wheel in new and worn rim profiles, respectively. The P-33 wheel was also found to reach a tread temperature of approximately 17% and 7% higher than the British wheel in new and worn rim profiles, respectively, for the thermal load associated with the Perth-Inverness drag brake case (Thermal Case 1). The P-33 wheel has an acceptable margin of safety for the two high-cycle (1×10^5 cycles and 1×10^6 cycles) mechanical fatigue loads and also for a 160,000 pounds mechanical load representing the dynamic effects of wheel flats. Thermal loads for the P-33 stop brake case (Thermal Case 2) are 30% and 50% lower for new and worn profiles than for the P-33 drag brake case (Thermal Case 1). Thermal stresses were found to be 17% and 7% lower for the P-33 wheel than the British wheel in the new and worn rim profiles, respectively.

CONCLUDING REMARKS

Although the North American wheels and axles discussed in this paper have historically performed well under generally heavier axle loads than those encountered in the United Kingdom, there are significant differences in railroad service and operating methods. Therefore, the types of computer analyses and design scrutiny described in this paper were warranted.

Analytical procedures for determining temperatures and elastic stresses produced by the expected mechanical and thermal service loads have proved useful in evaluating North American designs with respect to proven British designs. These analyses have provided information which enabled fatigue analysis, the determination of thermal lateral deflection in wheel rims, and the forces required to move the wheel on the axle. The J-36 wheel and class F axles are now in use on covered coil cars manufactured by Thrall Europa in York and operated by the English, Welsh and Scottish Railway.

Based upon our description of the service environment, and a review of the wheel removal data, it is clear that the North American environment is a severe test for railroad wheels. The long and successful track record of North American wheels in such a demanding environment should provide reassurance to customers around the world that our wheels are more than adequate to the task of freight service in the UK.

Analytical work on the 33 in. diameter freight car wheel is continuing. Efforts are under way to more accurately assess loading criteria and to determine appropriate wheel set designs for use in low-deck cars.

REFERENCES

1. Holley, Mel and Nigel Harris, "Britain's Rail Freight Revolution." *Trains*. Volume 58. Number 7. July, 1998.

2. Bridges, David. English Welsh & Scottish Railway, personal communication to the authors, February 4, 1999.

3. Association of American Railroads, "Rule 41." In *1999 Field Manual of the AAR Interchange Rules*. AAR: Pueblo, CO. 1999.

4. Association of American Railroads, "Freight Car Brake Design Requirements." In *AAR Standard 401,* AAR: Pueblo, CO. 1998.

5. Association of American Railroads, Car Repair Billing Data. AAR: Pueblo, CO.

6. British Railways Board, "Bass 504 Design Guide for Calculation of Stresses in Non-Drive Axles," August, 1985.

7. Association of American Railroads, "S-660 Procedure for the Analytic Evaluation of Locomotive and Freight Car Wheel Designs." In *AAR Manual of Standards and Recommended Practices-Section* G. AAR: Pueblo, CO.

8. Rusin, Thomas, "Thermal and Thermal/Mechanical Elastic Finite Element Stress Analysis of a J-36 S-Plate and a 952.5 mm British Railway Wheel," Rusin Consulting Corporation: Hasbrouck Heights, NJ. September 25, 1997.

9. Rusin, Thomas, "Thermal and Thermal/Mechanical Elastic Finite Element Stress Analysis of a J-36 S-Plate Wheel," Rusin Consulting Corporation: Hasbrouck Heights, NJ. September 26, 1998.

10. Rusin, Thomas, "Thermal and Thermal/Mechanical Elastic Finite Element Stress Analysis of a J-36 S-Plate Wheel," Rusin Consulting Corporation: Hasbrouck Heights, NJ. January 31, 1999.

11. Railroad Friction Products Corporation, "Dynamometer Test Plan 97-22 Thrall/EWS Perth Inverness Grade Simulation," July 31, 1997.

12. Rusin, Thomas, "Thermal and Thermal/Mechanical Elastic Finite Element Stress Analysis of a P-33 S-Plate Wheel," Rusin Consulting Corporation: Hasbrouck Heights, NJ. November 6, 1998.

Table 1

Chemical & Mechanical Requirements

Chemical Composition, percent.	BS5892 Part 3, R8T	AAR M107 Class B
Carbon	0.56 Max.	0.57 - 0.67
Manganese	0.80 Max.	0.60 - 0.85
Phosphorous	0.040 Max.	0.050 Max.
Sulfur	0.040 Max.	0.050 Max.
Silicon	0.40 Max.	0.15 Min.

Mechanical Properties	BS5892 Part 3, R8T	AAR M107 Class B
Rim Tensile Strength	860 - 980 N/mm^2	Not Specified *980N/mm^2
% Elongation	13% Min.	Not Specified *13%
Charpy U Impact, 20°C	15J, Min.	Not Specified
Hardness, Brinell	255 - 285 HBN	277 - 341 HBN

* Typical AAR M107 Class B Properties

Table 2
Three Brake Condition Cases

Maximum Effective Stresses (ksi), Temperatures (°F) & Deflections (in)
J-36 Wheel

$V_1 + L_1 + Th +$ Interference Loading

		60 mph (96.6 km/hr) Drag Brake *Case 1*	75 mph (120.7 km/hr) Stop *Case 2*	75 mph (120.7 km/hr) Drag Brake *Case 3*
Maximum Stress	*New Rim*	72.3 $\left(498.5\frac{N}{mm^2}\right)$	61.9 $\left(426.8\frac{N}{mm^2}\right)$	84.3 $\left(581.2\frac{N}{mm^2}\right)$
	Location	BRP	Tread	BRP
	Worn Rim	100.8 $\left(695.0\frac{N}{mm^2}\right)$	62.5 $\left(430.9\frac{N}{mm^2}\right)$	118.5 $\left(817.0\frac{N}{mm^2}\right)$
	Location	FHP	Tread	FHP
Maximum Temperature	*New Rim*	720°F (382°C)	460°F (238°C)	816°F (435°C)
	Worn Rim	975°F (524°C)	490°F (254°C)	1107°F (597°C)
Thermal Lateral Deflection	*New Rim*	0.0988 (2.51mm)	0.0253 (0.64mm)	0.1162 (2.95mm)
	Worn Rim	0.1106 (2.81mm)	0.0410 (1.04mm)	0.1761 (4.47mm)

Legend:
Vertical Load (V_1) = 56202 lbs (250kN)
Lateral Load (L_1) = 29225 lbs (130kN)
BRP = Back Rim Plate Fillet Region
FHP = Front Hub Plate Fillet Region

Table 3

Maximum Effective Stresses (ksi), Temperatures (°F) & Deflections (in)
J-36 Wheel and British Straight Plate Wheel

$V_1 + L_1 + Th +$ Interference Loading

		60 mph (96.6 km/hr) Perth Inverness Grade	
		J-36	*952.5mm British*
Maximum Stress	*New Rim*	$72.3\left(498.5\frac{N}{mm^2}\right)$	$136.7\left(942.5\frac{N}{mm^2}\right)$
	Location	BRP	FHP
	Worn Rim	$100.8\left(695.0\frac{N}{mm^2}\right)$	$189.0\left(1301.1\frac{N}{mm^2}\right)$
	Location	FHP	FHP
Maximum Temperature	*New Rim*	720°F (382°C)	647°F (342°C)
	Worn Rim	975°F (524°C)	883°F (445°C)
Thermal Lateral Deflection	*New Rim*	0.0988 (2.51mm)	0.1010 (2.57mm)
	Worn Rim	0.1106 (2.81mm)	0.1433 (3.64mm)

Legend:
Vertical Load (V_1) = 56202 lbs (250kN)
Lateral Load (L_1) = 29225 lbs (130kN)
BRP = Back Rim Plate Fillet Region
FHP = Front Hub Plate Fillet Region

Table 4

Maximum Effective Stresses (ksi) and Temperatures (°F)
P-33 Wheel

$V_1 + L_1 + Th +$ Interference Loading

		Thermal Load on Tread Location (ksi)	Maximum Temperature Tread(°F)
Rim Thickness	New Rim, $2\frac{1}{2}in$ (63.5mm)	$100.4\left(692\frac{N}{mm^2}\right)$	762°F (405°C)
	$1\frac{3}{4}in$ (44.5mm)	$123.9\left(854\frac{N}{mm^2}\right)$	827°F (447°C)
	$1\frac{1}{2}in$ (38.1mm)	$134.1\left(925\frac{N}{mm^2}\right)$	946°F (508°C)
	$\frac{3}{4}in$ (19.1mm)	$177.0\left(1220\frac{N}{mm^2}\right)$	1219°F (659°C)

Legend:
Vertical Load (V_1) = 56202 lbs (250kN)
Lateral Load (L_1) = 29225 lbs (130kN)
Thermal Load (Th) = Drag Brake from 60 mph on Perth Inverness Grade

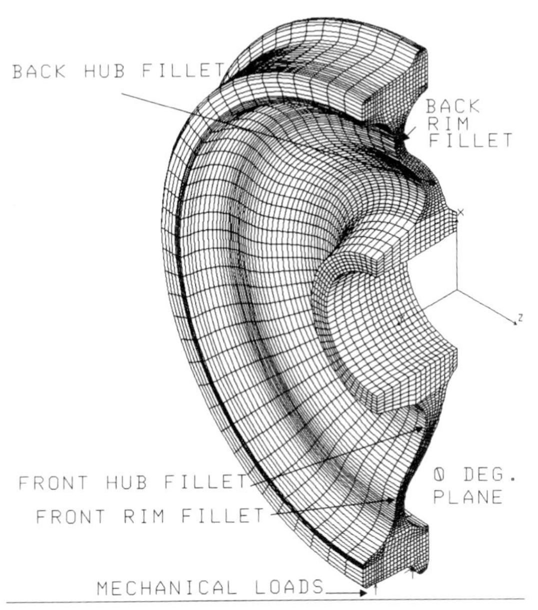

FIGURE 1 COORDINATE SYSTEM, POINTS OF LOADING AND PLANES OF REFERENCE

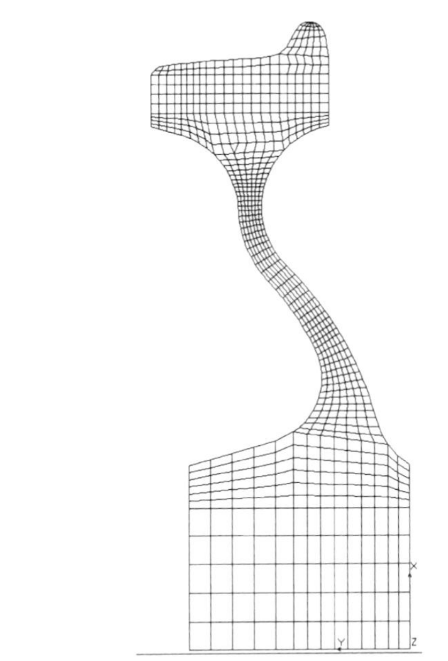

FIGURE 2 FINITE ELEMENT MODEL OF THE J-36 NEW RIM WHEEL

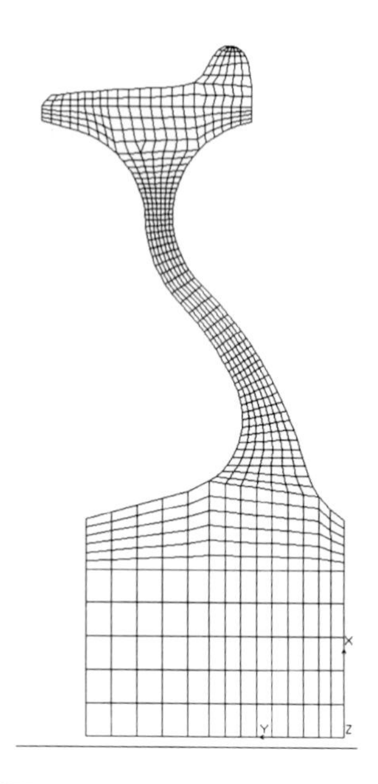

FIGURE 3 FINITE ELEMENT MODEL OF THE J-36 WORN RIM WHEEL

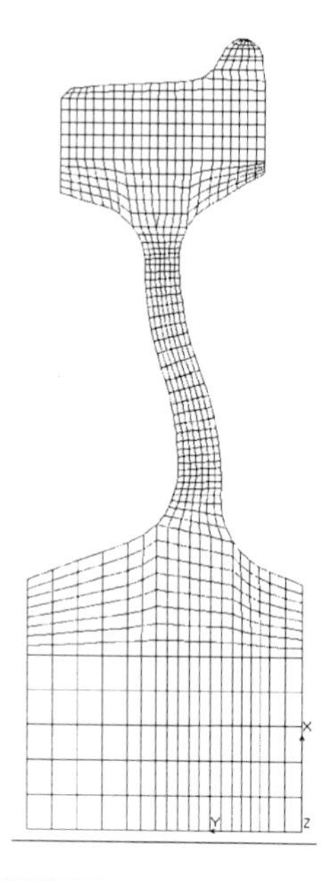

FIGURE 4 FINITE ELEMENT MODEL OF THE BRITISH RAILWAY NEW RIM WHEEL

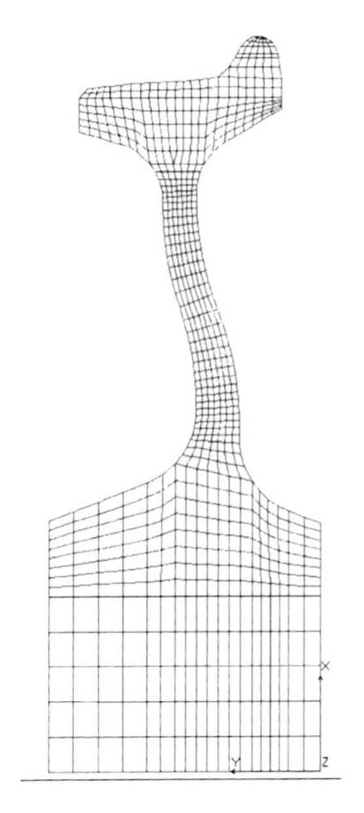

FIGURE 5 FINITE ELEMENT MODEL OF THE BRITISH RAILWAY WORN RIM WHEEL

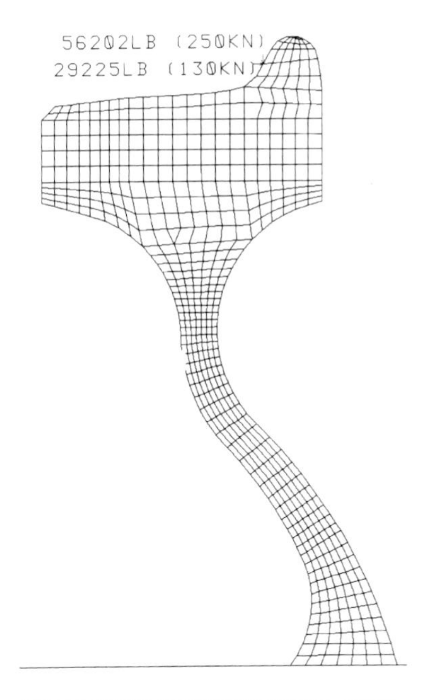

FIGURE 6 THE LOCATION OF THE V_1+L_1 LOAD

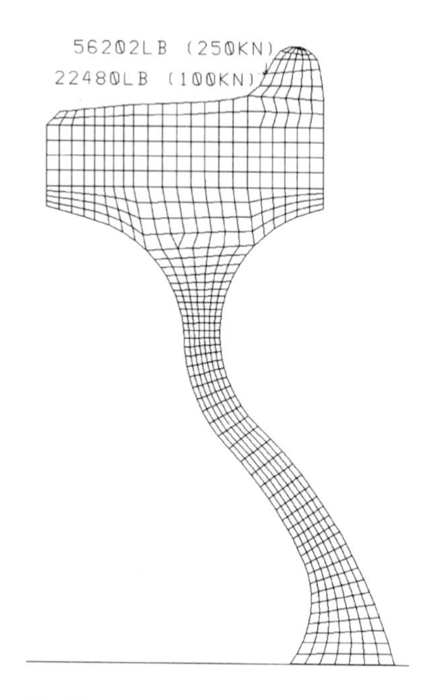

FIGURE 7 THE LOCATION OF THE V_1+L_2 LOAD

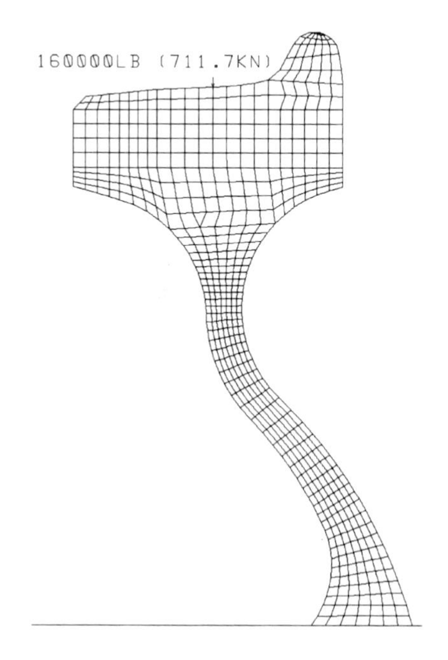

FIGURE 8 THE LOCATION OF THE V_2 LOAD

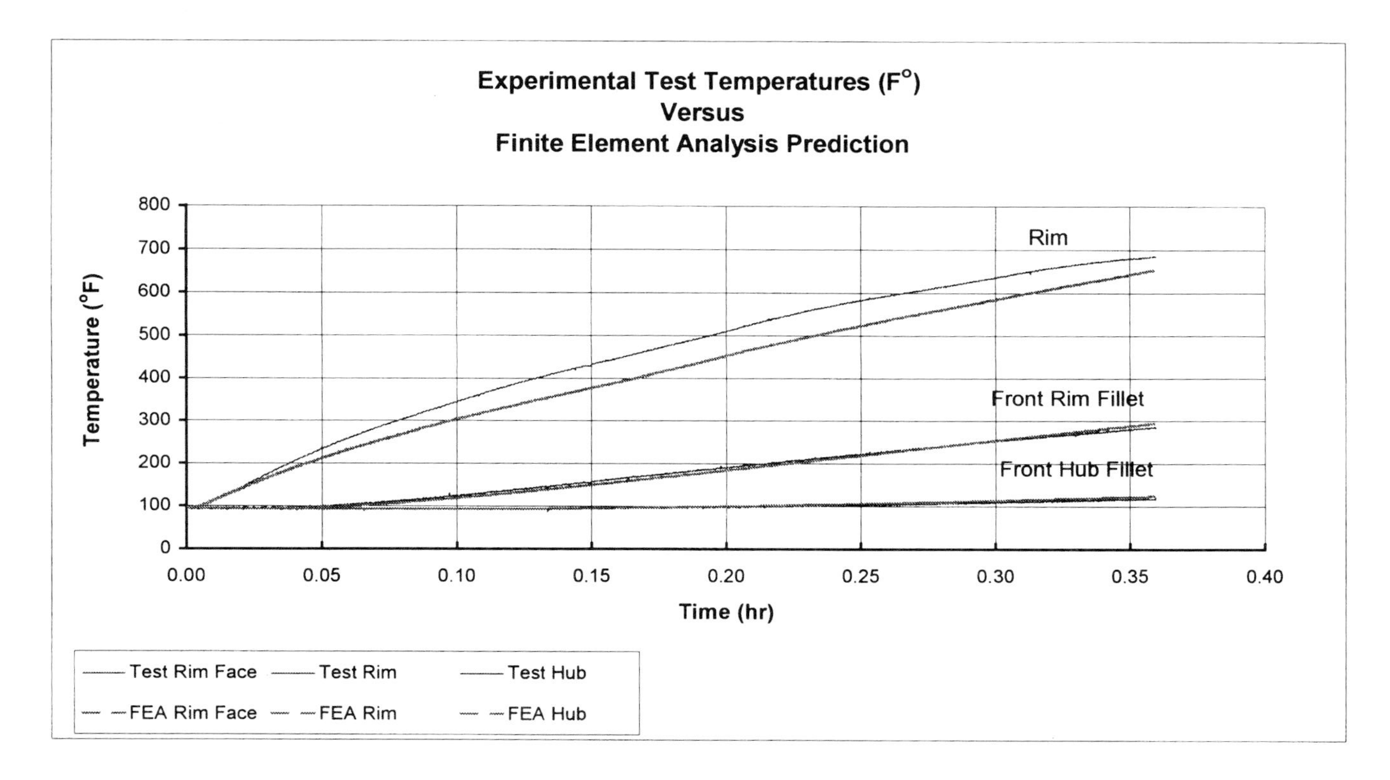

FIGURE 9 EXPERIMENTAL AND FINITE ELEMENT ANALYSIS THERMAL CORRELATION

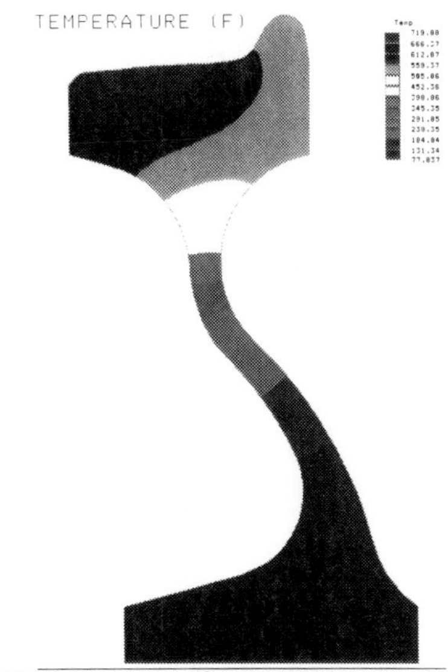

FIGURE 10 TEMPERATURE (F°) DISTRIBUTION IN THE J-36
NEW RIM WHEEL

FIGURE 11 TEMPERATURE (F°) DISTRIBUTION IN THE BRITISH
RAILWAY NEW RIM WHEEL

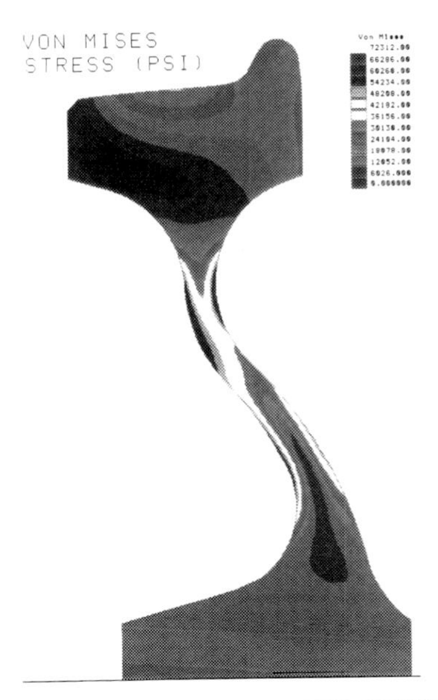

FIGURE 12 VON MISES STRESS (PSI) CONTOUR AT 90° IN THE J-36 NEW RIM WHEEL FOR THE V_1+L_1+THERMAL LOAD

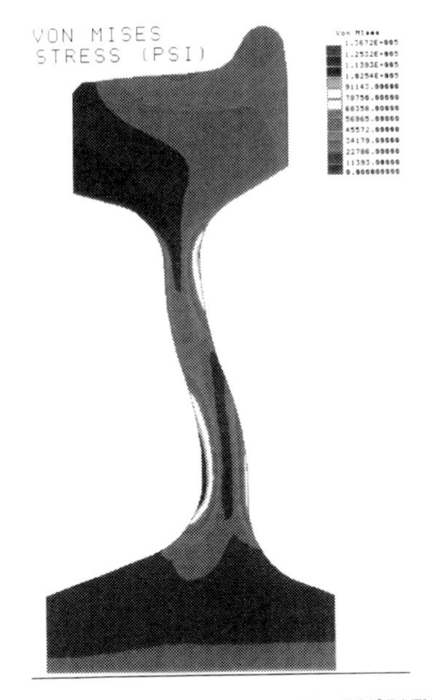

FIGURE 13 VON MISES STRESS (PSI) CONTOUR AT 20° IN THE BRITISH RAILWAY NEW RIM WHEEL FOR THE V_1+L_1+THERMAL LOAD

S629/006/99

Development of a low-carbon steel wheel for a disc-braked wheelset

F DEMILLY
Engineering Research and Development, Valdunes SAS, Dunkerque, France
B LOWE
Interfleet Technology Limited, Derby, UK

INTRODUCTION

By virtue of the criticality of the railway wheel it is not surprising that within Europe there have been no significant changes in the specified materials, for several decades. It is probably an inevitable situation as the approved grades of wrought steel defined in UIC and national standards do appear to be optimised for the majority of rail vehicle applications. That does not mean such materials give defect-free operation, rather that the user is reluctant to consider change with the associated increased safety risk, until experience has been gained.

This paper describes a break with tradition where a maintenance cost and effect situation was so onerous there was no serious alternative but the trial of a new material. That simple but radical change is described together with a précis explanation of the supporting investigation.

BACKGROUND AND THE DEFECT

In 1990, the Railfreight Distribution subsidiary of British Rail placed an order for 700 bogie container wagons. Based on the service history of an existing 20-year old fleet of 1500 first generation container wagons and a life-cost analysis, the evidence at that time was clearly in favour of a similar disc-only brake system for the new design. The known service weaknesses of the existing fleet were addressed in the new wagon based on a computer aided body design and UIC Y25 type bogies (VNHI-DA as UK version), Figure 1. Except for the state-of-art axle-mounted disc brakes, the wagon was UIC compliant to minimise the costs of design and manufacture for operation at 20.5 tonne axleload and 75 mile/h.

The subject of the paper, wheel tread damage, was not considered to be a significant risk. On the existing wagons, a wheelset that had to be changed outside overhaul was usually as a result of occasional wheel flats, brake defects or operational errors. The wheelsets would normally run 300,000km before reprofiling, the wheel tread exhibiting a highly polished surface. Confidence was also taken from several thousand other disc-braked wagons in the UK that were generally free of wheel thermal damage defects.

Following type-tests of structure, bogie and brake performance, deliveries of the new wagons into service commenced September 1991 through to December 1993. By February 1992, the first incident of wheel tread damage had occurred, escalating over the following months that by Spring 1993 a major problem was emerging. The work cost and content in reprofiling the wheelsets and loss of wheel life became the dominant element in the wagon availability and maintenance. The accumulated distance at which wheels had to be reprofiled for damage was random. From a minimum of 40,000km, the majority fell within a range of 80,000 to 150,000km, a long way short of the objective of compatibility with bogie maintenance periodicity.

The defect initially appears as areas of martensite (thermally transformed material) on the wheel tread surface, varying from isolated patches to a continuous band around the wheel.

The martensite is clearly evident as typified in Figure 2 with its trademark of a bright surface against the dull background of the remainder of the tread running surface. It invariably occurs within 70 to 80mm of the inside face of the wheel. Traces of martensite can also be found at the flange root.

The area of martensite is characteristically brittle and much harder than the original ferritic pearlite of the wheel from which it transformed. The continued running of the wheel generates very high cyclic stresses in the martensite with characteristic cavities developing by rolling contact fatigue over the subsequent 4-12 weeks. The cavities were up to 7mm deep with fatigue cracks penetrating a further 2-3mm. A common feature was that the tread damage was often asymmetric between the two wheels of the wheelset. There was also evidence of asymmetric damage occurring on the wheels of the wagon clearly indicating a track-related element of the damage on curves, transitions and switches. Such variability of wheel damage is not unusual on rail vehicles and was not considered significant.

The defect was not assessed as a serious safety risk as there was no evidence from similar damage on other vehicle types that it would cause a total wheel failure, a situation aided by the beneficial residual stress in the wheel. That judgement was also supported by a literature search. However, the loss of wheel life through extensive tread cracking, and potential bearing damage necessitated a closely managed regime by reprofiling as soon as damage was noted. It was achieved by undertaking maintenance at only one facility. The rate of occurrence of damage continued at a high level with clear seasonal variations confirming an influence of rail/wheel adhesion, Figure 3.

The 0.56%C wheel steel specified by British Rail for the application was BS 582 Part 3 Grade R8T, a requirement continuing today into UIC and Railway Group Standards. Wheel tread thermal damage is of course not new and has been the subject of many past studies and papers. What made this situation unique was that the incidence of damage reached a rate not anticipated or previously experienced on this or any other types of UK rolling stock.

INVESTIGATION

The high rate of wheel tread defects necessarily demanded an investigation to address:-

- what was the initiating cause of the defect;
- what features of the wagon were deficient despite the 'in-built' standardisation;
- what was different about the design compared to other wagons in the same traffic;
- what component interactive effects were resulting in the abnormal performance.

The detail is not relevant to the paper but the following illustrates the extent of its scope:-

- continuous maintenance surveillance and data analysis, and train drivers questionnaire;
- static tests on the brake system and a design review by the wagon supplier;
- repeat of the dynamometer disc-brake tests, friction evaluation of new/worn brake pads;
- repeat ride quality tests, tare/laden to evaluate P5, P6 and Pl0 wheel profiles;
- repeat on-track tests of brake performance and air system pressure stability;
- assessment of a wheel slide prevention system (WSP) and/or tread conditioning brakes;
- design review by SNCF (French Railways), commissioned by wagon supplier;
- theoretical study (VAMPIRE program) of bogie dynamics, focussing on wheel/rail contact area, the friction damped Y25 bogie is complex to model accurately.

None of these studies identified specific defects in the design, manufacture or maintenance that could be directly associated with the thermal damage. The only apparent differences between new and existing wagons was the 3-piece type bogie of the latter which has low stability wheelsets, and other disc-braked wagons which have a lower maximum speed at 60 mile/h. A WSP system would have helped reduce the damage incidence but the development time and costs were not favourable.

There was, however, a significant result from the theoretical study which supported the view that the defect initiation was within the wheel material and wheel to rail interface. The internal bogie forces during a brake application were predicted to stabilise the wheelset with rail contact at the flange root, which could give sufficient time for the thermal damage to occur in conditions of limiting adhesion, at the flange root and in a narrow band on the opposite wheel, (as the damage occurs in service). This wheelset behaviour was also seen during the second series of track tests particularly when braking from higher speeds.

The study also suggested that as the wheel profile wears, the bogie dynamic behaviour was more likely to result in increased movement of the wheel across the rail (instability/hunting) and may prevent the build-up of martensite in the characteristic narrow band such that wheels could run to a worn profile without damage. Unfortunately, due to the low wear rate of the disc-braked wheel, that would only benefit a small percentage of the wheel population as the thermal damage was so dominant. There is some recent evidence of this from maintenance surveillance with a reduction of damage as wheelset instability increases with normal bogie suspension wear. Intentional de-stabilising the wheelsets as a design criteria was not an option.

DEFECT ORIGIN

The characteristics of the tread defect have remained consistent throughout the wagon fleet since first identified. The defect was clearly thermal damage in origin as a consequence of wheelslip, principally during braking from high speed (70+mile/h), in adhesion conditions below the limiting value required by the wagon brake demand.

The mechanism of one theory is well documented that the wheelslip energy causes instantaneous heating of the wheel/rail contact area and a transformation of the metallurgical structure in the surface layer. As the wheeltread rotates away from contact with the rail, almost instantaneous cooling is provided by the wheel rim mass. The local quenching can create martensite and this quenched structure receives no tempering thus increasing its susceptibility to cracking (1). Repeated wheelslip events can build-up and extend the area of martensite.

With continued wheel rotation, rolling contact fatigue generates cracks in this very brittle structure between the basic steel structure and the martensite. This effect is also amplified by the differences in thermal expansion of martensite and its 0.5% larger volume compared to the original ferritic-pearlitic structure, and by the fact the martensite hardness is directly linked with the carbon content (2). An example of a martensite white spot and fatigue crack is shown in Figure 4. This hypothesis means that the tread layer temperature in the wheel/rail area can reach values more than 750^{0}C. A hardness survey was made of typical areas of martensite on the tread. As-manufactured tread values of 295 Hv had work hardened typically to 330 Hv, with the thermally transformed areas typically 450/600 Hv.

A second theory considers the influence of the loading on the phase diagram transformation. The pressure, in the contact area, can decrease the austenite start temperature (Ac3) and martensite start temperature (Ms) by approximately 80ºC for a pressure of 25kbar (3), and modify the Isothermal Transformation diagram limits. Under plastic deformations, cementite decomposition is possible at low temperature (4), less than 400ºC, by a carbon migration in ferrite dislocations. So, a structure quite similar to the martensite appears with a high deformation level. Two mechanical parameters also seem to have a great influence on the phenomena: the mechanical load frequency and the shear stresses.

A POTENTIAL SOLUTION

In parallel with the engineering investigations, an extensive metallurgical investigation had pursued the possible benefits of a wheel material, different to that specified but traditionally used without question. The subject has been examined in past papers for more than 15 years and particularly at the periodic International Wheelset Congress. Independent studies had been made in early 1980's by BR Research (5) and SNCF Research/Valdunes (6), amongst others, to find a wheel steel of reduced susceptibility to the thermal damage. The BR development, in part with the British Steel Corporation, was consequential to work on bainitic fracture-tough rails. Laboratory trials were made with 0.10%C steel but never progressed to vehicle applications. The SNCF/Valdunes development to find a steel with combined R6 and R8 characteristics produced two grades with 0.37 and 0.25%C which had limited vehicle trials.

Metallurgical testing:
The high level of damage incidence initially suggested a possible material fault, but investigations by the wheel supplier and an independent specialist both concluded there were no manufacturing defects. The steel composition, cleanliness, structure and mechanical properties were correct. The Rm strength level of batches of R8T wheels were between 880 Mpa and 945 Mpa. The values of the majority of wheels in service were located in the lower part of the specification range in BS 5892 part 3 with an average Rm at 908 Mpa and Re yield limit of 605 Mpa. Residual stress measurement using the ultrasonic birefringence method gave a level of -100 to -120 Mpa which can be considered as a normal compressive level on such wheels in service. No evidence of overheating was detected.

Metallurgical solutions:
In considering metallurgical possibilities, it is easy to establish that the main railway steels are plain carbon, and it is well known by the rail industry that thermal crack defects are directly associated with carbon content and martensite. This can be quantified by the martensite start temperature that can be linked to steel grade composition with an analytic formula. Different equations have been proposed to describe Ms°C temperature. Using the HOLLOMON formula (7) of SNCF, it shows the main parameter of Ms°C is the carbon content. The same study, investigating cracking and fractures of railway wheels, had shown experimentally that with a temperature over 285°C for Ms°C, it is possible to limit wheel tread defects.

$$Ms°C = 500-350(\%C)-40(\%Mn)-35(\%V)-20(\%Cr)-17(\%Ni)-10(Cu\%)-10(Mo\%)+30(\%Al)$$

Martensite formation also closely depends on the heating conditions. A high temperature heating of the steel will give good alloying element dissolution and will have a positive effect on the martensite start, in term of tread defects. Instantaneous temperatures obtained during wheelslip and the braking time are quite high compared to standard heat treatment. It means that this heating cycle favours martensite formation.

Metallurgical optimisation:
Considering Ms°C and the phase transformation point Ac3°C as criteria, it is observed that the main wheel steel grades have a martensite start temperature below 285°C for UIC grades and below 250°C for standard AAR steel grade (class B and C). Much higher values can be found for example for 28 CDV 5.08 steel grade, with vanadium additions used for high-speed train brake disc applications. However, this grade with Rm in a range of 1050/1250 Mpa, for a Re yield limit over 970 Mpa has no wheel applications.

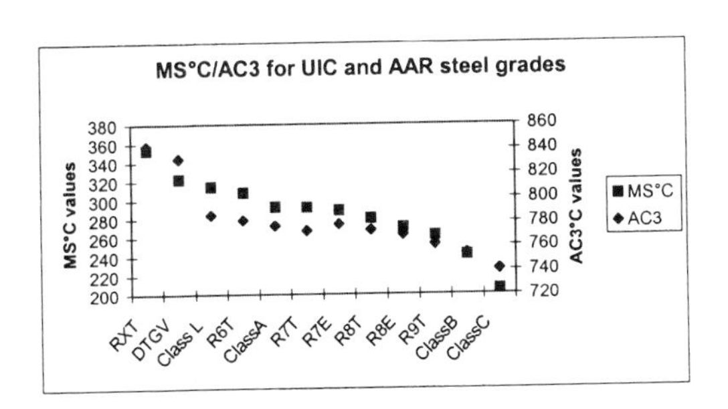

With the HOLLOMON formula, the following variations are found depending on the steel composition. Carbon gives a continuous decreasing effect on the Ms°C temperature. A difference of less than 0.1% weight of carbon (it means the difference between maximum carbon content of the R7T and R8T steel grade) gives a decrease of 28°C. Manganese gives the same effect with a decrease of 6°C for a difference of 0.16% weight. The cumulative effect of Chromium and Nickel (0.24 % weight) gives a decrease of 9°C.

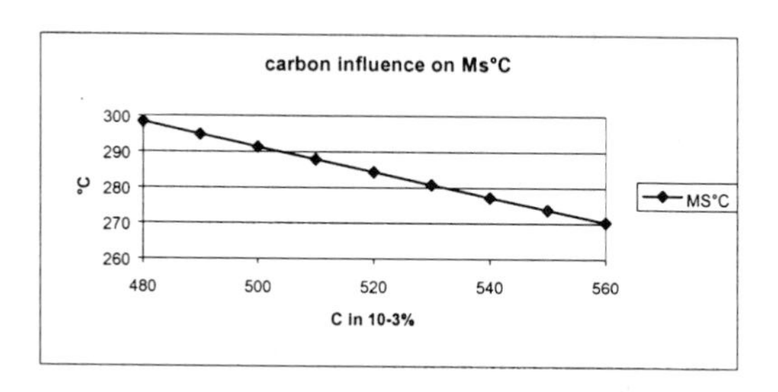

Manganese also gives a negative effect on the Ac3°C temperature start in the same range as for the carbon content. But this effect can be limited by addition of alpha-phase forming elements such as Silicon, Chromium and Molybdenum.

In summary, it was concluded there was potential to use a low carbon content steel. With this low carbon content and in order to maintain the same hardenability, micro alloying additions were added to obtain the specification steel strength. Hardenability can be determined using Jominy curves. With a low carbon steel (LC) grade (0.22%) and the residual element of the R8T steel grade, it can be seen the hardness level is much lower than the specification limits. Additions of elements as Manganese, Silicon, Chromium and Molybdenum give the higher curve (8).

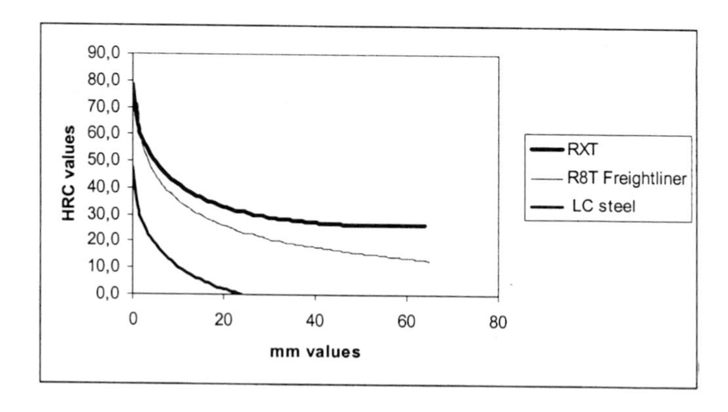

PROPOSED LOW CARBON STEEL

To perform a service trial in the same conditions as the existing wheels and minimise risk at all stages, a formal specification was issued and the material identified as RXT, particular to the application. The specification required the wheels to accord with all aspects of BS 5892 Part 3 Grade R8T through manufacture and tests, except as varied by the chemical composition. The developed specification restricted the carbon to a maximum of 0.25%. It was recognised there would still be the risk of thermal damage but it was opined that any martensite formed would be thin, relatively soft by virtue of the low carbon content and wear away before generating shells or cavities. A bainitic steel of less than 0.1%C was considered too radical as a single step change, being relatively unknown for the application and at risk of poor wear performance.

Chemical analysis:

10^{-3}%	C	Mn	Ni	Cr	Mo	V	Al	Cu
R8T	530	700	120	130	40	0	20	140
LC steel	220	700	120	130	40	0	20	140
RXT	220	>700	<300	>300	<300	0	20	140

After forging, wheels were heat treated, austenitisation at 850°C, water-quenched and tempered during 3 hours at more than 500°C, the mechanical properties obtained following specification BS 5892 Part 3:1992 were.

	Ultimate limit Rm	Yield limit Re	A%	KU+20°C
RXT Rim	**886 Mpa**	**741 Mpa**	**17**	**20J**
RXT Web	**817 Mpa**	**622 Mpa**	**17**	
BS 5892 Part 3	**860/980 Mpa**	**>600 Mpa**	**>15**	**>17**
R8T Average	**908 Mpa**	**605 Mpa**	**16**	**19**

The strength limit of the RXT is located in the lower part of the R8T distribution, and is 22 Mpa under the R8T average value. Nevertheless, RXT has a rim yield limit of 21% higher than R8T carbon steel and 38% in the web which is also a valuable feature for the wheel design. The difference in tensile strength of RXT between the rim and the web is about 69 Mpa compared with a difference of about 200 Mpa for the carbon steel, which is normal due to the alloying additions.

Hardness homogeneity had the same tendency with a range of 277/285 HB for the wear volume and 269 HB near the web.

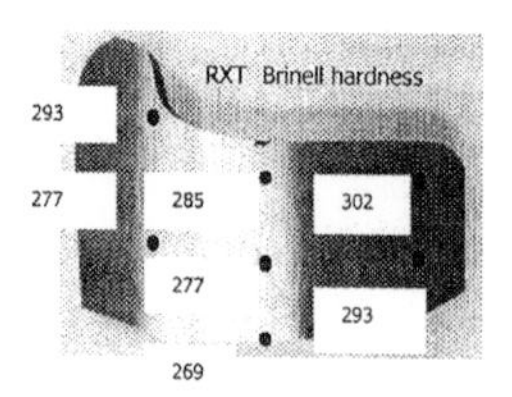

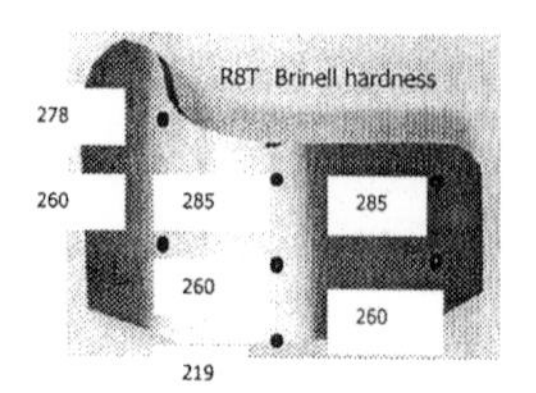

RXT - Brinell Hardness Measurements – R8T

There are also differences between hardness measurements and strength values due to the alloying steel grade. The hardness difference between the new and worn rim is 25 HB and 66 HB between the maximum tread value and the limit at the rim/web interface. The gap is lower on the RXT with a difference of 8HB on the rim and 16 HB between the rim and the web. This can also explain the residual stress level after quenching of about 100 Mpa. The alloying steel grades have lower residual stress values than carbon steel grade.

The RXT micro-structure is bainitic tempered in the rim area compared to the classic ferritic-pearlitic for the R8T standard steel grade. In each case the grain is fine and homogenous, see Figure 5.

Concerning fracture toughness, measurements following UIC 812.3 have given:

Steel grade	C	Strength	Average KQ
	10-3%	Mpa	MpaVm
R8T	530	908	87
RXT	220	886	98

There was no perceived requirement to measure the intrinsic fatigue strength of RXT on the basis of it having the characteristics of a low carbon alloy steel entirely suited to a railway wheel.

SERVICE TRIAL

A pre-requisite to the possible future approval of the new material was a managed trial to demonstrate its performance and that there was no identifiable safety risk. The long-term objective was that a successful trial would lead to use low-carbon wheels for fleet replacement.

The experiment batch of 40 Grade RXT wheels manufactured by Valdunes were fitted into 6 container wagons in November 1994. The wagons were operated without restriction in the UK network services of British Rail and then Freightliner Ltd who became the operator of the wagons in 1996. The management of the trial included a detailed surveillance of wheel performance at the standard maintenance events. To ensure consistency of reporting, the examinations were made by the same personnel and included sampling by a metallurgist.

Trial Summary:
The monitoring of wheels is continuing, but at March 1999 it was possible to conclude that the principal objective had been attained by the low carbon steel RXT with the elimination of the majority of thermal damage, compared to the standard R8T steel. The only martensite seen was small isolated patches and thought unlikely to develop into cavities being thin and relatively soft, figure 6. It is predicted that the achievable wheel profile life will be at least 300,000km and a forecast average of 400,000-450,000km; a factor of 3 improvement on the standard wheel. That immediately provided obvious wheel life cost benefits. There are additional benefits through minimising the abnormal wear on the wagons through damaged wheel vibration inputs, extension of maintenance periodicity and improved wagon availability.

Trial Continuation:
It was never anticipated or claimed that RXT would solve all the wheel tread problems that the fleet of wagons may encounter. Indeed it is possible that other types of defect may occur at long profile life, previously disguised by the thermal damage on R8T or peculiar to RXT material. They may be the limiting criteria rather than the profile wear rate.

- Profile wear – the sample size of RXT wheels has not provided sufficient data to conclude on the wear rate but it does appear similar to R8T, on both tread and flange. There is no evidence of susceptibility to rail head debris damage. A longer term surveillance will be necessary to identify any tendency to wear out-of-round or multiple flats due to wheel skip over rail discontinuities.

- Wheel flats – with no WSP on the wagon, these will inevitably continue to occur (as on similar wagons) when adhesion conditions are adverse; some of the RXT wheels in the trial have had to be re-profiled due to slide flats but there was no evidence of martensite at the damage sites.

- Martensite type flats, could be RXT specific – one of the trial wheelsets had some very thin, soft martensite which apparently wore or fractured away with time and may have created shallow flats eventually developing into damage through rolling contact fatigue.

- Rolling Contact Fatigue – at long profile life, some RXT wheels have a band or patches of RCF, at an oblique angle, on the tread at the change point of the profile slope. All wheels with such a profile transition are at risk by virtue of the stress concentration effect on the wheel/rail contact area. Tread-braked wheel wear rate is high enough to disguise the RCF; disc-braked R8T wheels exhibit RCF but it usually does not propagate unless associated with a wheelflat; the RXT wheel appears a little more susceptible to RCF than R8T but current experience is insufficient to state if it will limit profile life.

SUMMARY

What has been achieved at March 1999?

- It can be shown that RXT steel grade fulfils the requirements of the specifications

BS 5892 Part 3, 1992 and UIC 812.3, especially for mechanical properties and fracture toughness. Differences only exist for the metallurgical structure and chemical analysis. Other tests have been conducted by SNCF especially for integral wheels which are used as wheel and brake disc. Rig tests of thermal crack resistance have also confirmed the good quality of the material.

- It was always going to be difficult to identify the point at which the trial could be concluded – waiting for thermal damage not to occur. However, the results have been positive and Freightliner extended the trial in January 1998 by ordering 200 RXT wheels to cover their immediate need for replacement.

What further development is necessary?

- Despite the success of the trial it is freely acknowledged that RXT may not yet be optimised. Development of wheel materials clearly has a long gestation period and the trial of low carbon steel has raised a number of questions including:

  - what is the optimum carbon content – safety risk/tread damage risk/wear;
  - what are the optimum mechanical characteristics;
  - does thermal damage wear or degrade preferentially and sufficient to cause flats;
  - what chemical composition will give the best resistance to rolling contact fatigue;
  - would a new wheel profile minimise rolling contract fatigue but maintain wheelset dynamic performance;
  - how would a low carbon steel wheel perform in a vehicle with WSP.

What are the opportunities for other applications?

- It is believed this has been the first full trial with low carbon steel wheel that has progressed close to a conclusion. The results have been very encouraging. There is potential with this type of steel for other applications on disc-braked wheels with/without WSP.

- Grade RXT was developed in response to a specific problem and cannot be presumed the instant panacea for all similar applications. Each case should be carefully assessed through a monitored trial and its use justified through technical and risk assessment. As example, Freightliner have already commenced a trial on a different fleet of container wagons with a low carbon steel from another supplier.

ACKNOWLEDGEMENTS

The opinions expressed in this paper are those of the authors who acknowledge the metallurgical contributions of Mr D W A Ward, Serco Railtest Ltd, Mr J C Tourrade, SNCF and Mr B Catot, Valdunes SAS.

Acknowledgement is made to Freightliner Southampton Depot for their assistance with the trials.

The authors would also like to thank the Directors of Freightliner Ltd, Valdunes SAS, and Interfleet Technology Ltd for permission to publish this paper.

REFERENCES

1. "Theoretical and experimental study of wheel spalling in heavy haul hopper cars" D. Stone, Br. Rajkumar, Belport, Hawthorne, Moyar.
2. "Traitement thermique" Acier inox p21 n°14, 1999.
3. "Principes de bases des TTH" p126 A. Constant, G. Henry, Jc. Charbonnier.
4. "Phase blanche dans les rails" Note Irsid Mpm 97 N 1126, Bertrand, Guelton, Galtier, Scott, Seux, Juckum.
5. "Bainitic Steels for Railway Wheels" K.J. Sawley, E.G. Jones, J.A. Benyon, International Wheelset Congress 1988.
6. "New Steel Grade for Monobloc Wheels" B. Catot, J..C Tourrade, International Wheelset Congress 1985.
7. "Un choix de la SNCF": les roues monoblocs en acier non allié traitées en surface" Mp. Ravenet, Mp. Gautier Juin 1963.
8. "Les aciers spéciaux" p231 Beranger, Henry, Labbe, Soulignac.
9. "Martensite and Contact Fatigue Initiated Wheel Defects" E. Magel and J. Kalousek, International Wheelset Congress, September 1998.
10. "Identifying and Interpreting Railway Wheel Defects", E. Magel and J. Kalousek, IHNA Congress June 1996.

FIGURE 1 Y25 TYPE VNHI – DA BOGIE

FIGURE 2 R8T WHEEL TYPICAL THERMAL DAMAGE – AREAS OF MARTENSITE AND CAVITIES

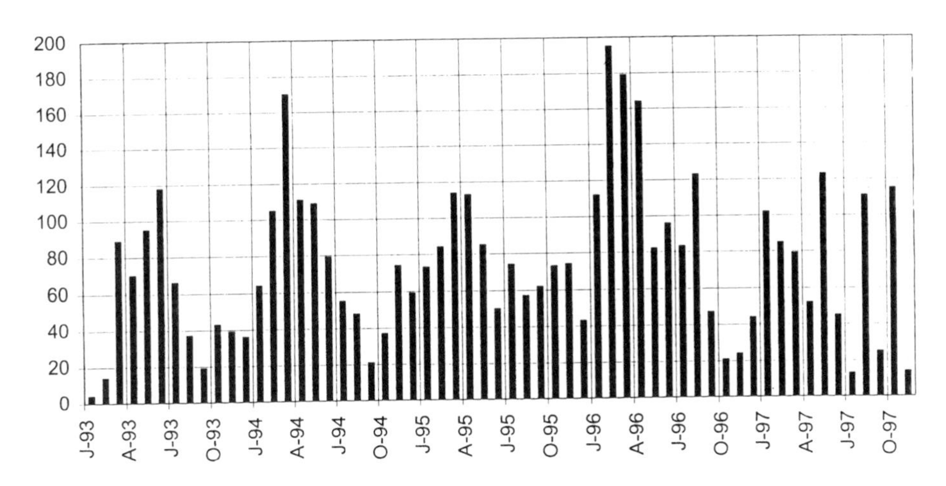

FIGURE 3 **FSA/FTA WAGON WHEELSETS RE-PROFILED DUE TO TREAD DAMAGE FROM JANUARY 1993**

FIGURE 4 R8T MARTENSITE WHITE SPOT WITH UNDERLYING FATIGUE CRACK

FIGURE 5 R8T FERRITIC-PEARLITIC STRUCTURE X 200

FIGURE 5 RXT BAINITIC STRUCTURE X 200

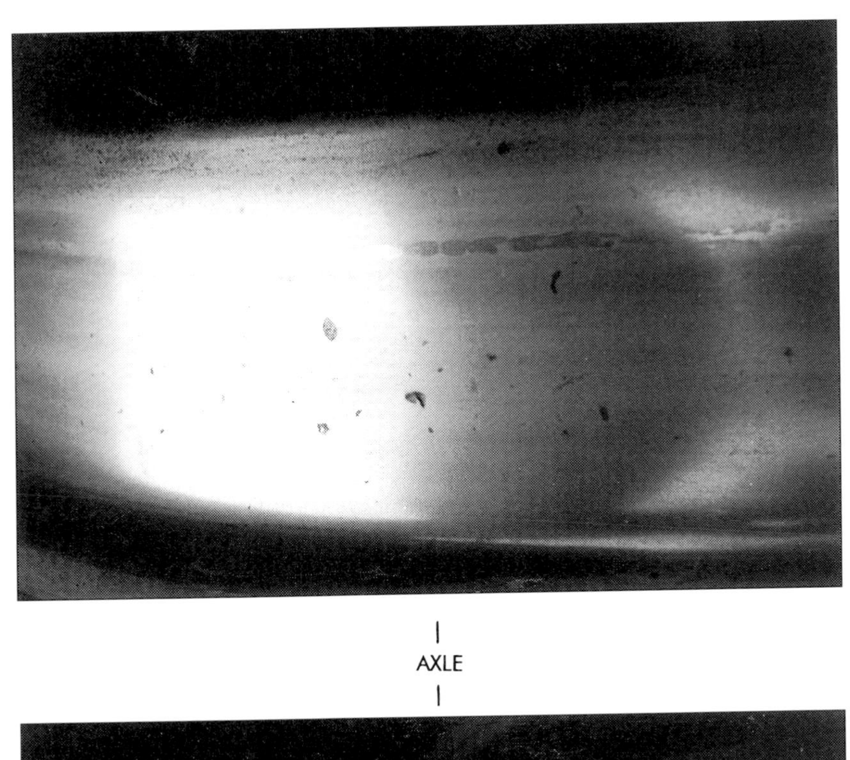

AXLE

FIGURE 6 RXT WHEELS AT 210,000KM ON SAME AXLE – SMALL AREAS OF SOFT MARTENSITE

S629/007/99

Cast wheels for operation in the UK

D BRIDGES
English, Welsh, and Scottish Railway, Nottingham, UK

At an early stage in the class 66 design scrutiny process, the class 59/2 wheel had been validated for the steeper gradients and higher braking duty expected of the class 66.

However, the manufacturer was at the time in some difficulty, and EWS were considering alternatives to ensure a secure supply.

EWS had for some time been investigating cast wagon wheels, but this exercise was broadened to include a locomotive wheel.

Initially a 40" (1016 mm) wheel was considered but this was rejected for 2 reasons:-

- Uncertainty that all private sidings met Railtrack gauge clearance to 75 mm ARL between rails, (gearcase and motor clearance).
- Stresses due to thermal input were 37% higher for the 40" compared to the 42" wheel.

Knowing the past problems with rim stress reversal, EWS were keen to procure a wheel with good thermal tolerance to avoid wheel fractures such as that which occurred on 60022 from a 0.5 mm deep defect.

Parabolic deep dish 42" cast wheels by the Griffin wheel company were selected as Griffin is the market leader for cast wheels in the United States.

The performance of these wheels was compared to the existing class 59/2 wheel work and information was gathered on failure rates, manufacturing processes and operational environment. Metallurgical and NDT aspects were addressed by an independent consultant.

Thermal Input

US locomotive wheels are designed to accept thermal loads in spite of the fact that in train, the locomotive brake is released except in an emergency. Also dynamic brakes are in common use where gradients are encountered. In order to obtain service experience of a generically similar wheel under braking loads, the power input to wagon wheels was evaluated. Even allowing for the low gradient descent speed, typically 20 mph, and dynamic brake, wheel power inputs can be high due to the high axleload and variance of the US triple valve pneumatic system used. Also, gradients can be longer and/or steeper compared to the UK, so energy inputs are significantly higher.

Wheel Type/ Wheel Size (mm)	Axleload (Tonnes)	Speed (mph)	Route	Input Power/ Wheel (kw)	Drag Losses (kw)	Braking Required (kw)	Wheel Input	Energy Input (MJ)
EWS 953 (37.1/2")	25.5	60	Perth Inverness	45	6	39	26	34
US 914 (36")	32.4	20	Cajon Pass	45	1 Drag 12 Dynamic Brake	32	21	83

Table 1

Assumed Gradients

Perth – Inverness 1/75 for 22 miles.
Cajon Pass 1/33 for 6 miles followed by 1/50 for 16 miles.

An interesting feature of the Cajon Pass gradient is that the friction brake power will remain relatively constant for the full 22 miles, in spite of the gradient easing.

This is because as the direct release friction brake cannot be released whilst still on the gradient, speed is maintained on the flatter section by reducing the dynamic brake. Therefore, power inputs are only slightly higher than in the UK and energy inputs are much higher.

It should be noted that the class 66 has a nominal axleload of 21 tonnes. Examination of braking reports with laden tare brake MGR wagons gave an average brake axleloading of 2t per axle. Thus from a braking perspective the axleload is effectively 23 tonnes.

To validate the drag assumptions used from specification CP-DDE 129, EWS performed a coasting test on the ECML with a 2500 tonne steel train hauled by a class 60. This showed close agreement and gave a value of
13 kw loss per 25.5 tonne axle at 60 mph. To coast down to 45 mph from 60, the train took 8 miles, an impressive demonstration of the efficiency of rail.

Interface Issues

In terms of interface issues, EWS were keen to use a standard US wheel width. Unlike wagons which have a standard rim width of 5.23/32" (145 mm) locomotive rim width options vary between 5.1/2 (140 mm) and 5.5/8" (143 mm). EWS choose the wider width as:-

- This was considered to offer a greater heat sink.
- It was desired for reasons of block life and standardisation to use the 3.3/8" (85 mm) US standard block width (standard UK 2.1/2" (64 mm)). The wider rim width reduces the risk of chamfer cracking.
- The class 66 uses conventional brake rigging due to the steering displacements, in place of the direct acting SAB BFC brake units used on classes 58, 59, 60 and 92. This arrangement has a greater risk of lateral block displacement and again a wider wheel compensates and reduces the risk of chamfer cracks.

Rim hot lateral displacements were also evaluated. Due to the offset of the rim relative to the hub connection, heating of the rim produces an elastic displacement toward the centre of the axle, which is reversed upon subsequent cooling.

Historical specifications such as the class 60 have required that the back to back dimension not to vary by more than ± 3.5 mm due to this effect.

The finite element model of the class 59/2 wheel was predicted to move 1.6 mm per wheel after an input of 20 kw for 15 minutes (Perth – Inverness at 75 mph).

The figures below are for the 42" cast wheels after heating by induction coil for the period shown.

Power Input (kw)	Duration (mins)	Deflection (mm) Toward Axle Centre Per Wheel
49	30	3.6
30	60	3.3
24	90	3.0

None of the above cases exactly replicates the Perth – Inverness heating assumption, but displacements are likely to be very similar to the class 59/2 wheel which are within the previous historical limit. EWS has separately scrutinised the previous ± 3.5 mm limit and concluded that it is conservative from a safety viewpoint, as regards contact with check rail entries and switch blades.

Material Properties

AAR Class C (0.67 – 0.77% Carbon) steel is in common use in the US as it is harder and more wear resistant than Class B (0.57 – 0.67 % Carbon). Studies show that the wear rate of Class C material is 70% lower than Class B. However, EWS considered Railtrack may object to a harder wheel and in any case, the fracture toughness of the Class C steel is much lower than Class B. EWS considered Class B to be a conservative choice which had broadly similar fracture toughness to AAR Class B (wrought) and R8T, and hardness close to R8T. The fracture toughness of the cast Class B material is superior to the UK BR 108 wheel steels manufactured until 1983.

Some properties are listed below.

	AAR Class B Cast	AAR Class B Wrought	BS5892 R8T
Carbon %	0.57 - 0.67	0.57 - 0.67	0.52 – 0.56
UTS (MPA)	972 – 1034	870 – 970	860 – 980
Yield (MPA)	620 – 758	480 – 600	540 – 600
Hardness (HB)	277 – 341	240 – 310	355 – 285
% Elongation	7%	Approx 13%	13%
Fracture Toughness $K_{1}C$ (mpa $\sqrt{m}$)	60 – 68	Approx 75	Approx 75

Manufacture

The Griffin wheel company have a patented "controlled pressure pouring" process, where the mould is filled from the bottom up from a pressurised vessel of molten metal. This results in less atmospheric contamination of the molten steel, and a sounder casting.

Graphite moulds are used which require reprofiling every 5 or 6 wheels to maintain accuracy and finish. The graphite moulds result in excellent balance, and wheel radial runout as cast is typically within 0.25 mm. Tread profiles can be "as cast", to the UK tread contour tolerance of $\pm$ 0.25 mm.

Due to the relatively thin section of the cast wheel compared to a forging ingot, hydrogen can readily defuse. Hydrogen content allowed is 3 PPM maximum.

All wheels are magnetic particle tested and rims are radially and transversely ultrasonic inspected. Wheels are also 100% shot peened for the generation of beneficial residual stresses.

Failure Statistics

There are approximately 1.2 million freight wagons in interchange service in the United Statbs, and 18,000 locomotives. The Griffin Wheel Company has a majority share of this market. The values shown below relate to wagons in interchange service have been independently verified by the Federal Railroad Administration.

	1992	1993	1994	1995
Derailment caused by wheels	79	77	49	55
Derailments caused by broken wheels	22	19	12	19
Derailments caused by broken Griffin wheels*	3	2	3	1

* Griffin internal data, FRA data does not distinguish between manufacturers.

Manufacturer Development

The following techniques among others are in use by the manufacturer:-

- Finite element modelling.
- Dynamometer testing.
- Induction coil heating and sawcut testing.

The latter method involves fitting an induction heating belt around the wheel tread and subjecting the wheel to varying power input levels for different durations. After heating the rim is sawcut radially to a depth of 5”
(127 mm) and displacement across the cut is measured and plotted relative to cut depth. Multiple results can then be used to build up a graph of heat input/duration and the effect on rim bulk residual stresses.

EWS Testing/Monitoring

- EWS has validated the resistance formula given in specification CP-DDE-129.

- EWS has performed brake tests at 75 mph involving the descent of the WCML Beattock gradient (1/75 for 10 miles) with a 1700t train of laden ex National Power hopper wagons. The train was brought to a stand at the foot of Beattock and temperatures were measured using an infra red camera. These were lower than expected, the maximum bulk rim temperature being 93°C. This was estimated to result from an average power input over and minutes of 26.8 kw. FEA modelling with this thermal load indicated a maximum thermal stress of 140 Mpa in the outside of the web. Following cooling no significant residual stresses were predicted in the web or rim. It is possible that this test is not the worst case, as the train was formed of modern wagons with load proportion braking. EWS are planning a further test with tare braked Merry Go Round wagons in the laden condition on the Onllwyn branch line in South Wales. This line is 10 miles long at an average gradient of 1/70. Trainspeed is typically 20-25 mph giving a heating time of about 30 minutes. Due to the speed, power inputs will tend to be lower, and there will be more on-loading of the locomotive brake.

Summary

- Good geometry with low stress levels.
- US has comparable power inputs to UK on wheels of same generic type.
- Interface issues addressed.
- Good material properties, fracture toughness close to AAR Class B (wrought), and R8T, better than pre BS5892 wheel steels.

S629/008/99

New NDT techniques for railway wheels and axles

Y SOOR
Angel Train Contracts Limited, London, UK

1. INTRODUCTION

Following the accident at Rickerscote near Stafford in March 1996, where an axle fractured on a freight wagon wheelset resulting in a fatality, a report was commissioned into its failure, as part of the HMRI investigations. Included in the report was a recommendation that NDT methods which are used in other industries should be investigated for possible use in the rail industry. One technique specifically mentioned in the report was Alternating Current Field Measurement (ACFM), which this paper attempts to cover, along with another technique. Therefore, following on from the recommendations of the report, the Rickerscote Steering Committee commissioned a study to investigate ACFM, and similar NDT techniques to determine:

1. Whether they would be suitable for the rail industry
2. What developments of them would be required
3. The costs involved and the next steps

Angel Train Contracts Limited representing all the ROSCOs and English Welsh and Scottish Railways (EWS) as part of the Steering Group, were tasked with developing the brief for the study, and preparing the Invitation to Tender. After tender evaluation the study was awarded to SERCo Railtest, and although this study is not yet complete, all the base data has been collected, the findings of which are the basis for this paper.

1.1 Scope

The new techniques that this paper presents are Alternating Current Field Measurement (ACFM), as developed by Technical Software Consultants and Field Gradient Electro Magnetic Array (EMA), similarly developed by ALSTOM/NEWT. The paper starts by covering the issues surrounding current NDT methods, and the importance of Fracture Mechanics as a prediction tool. Thereafter, the new techniques are described and then, after concluding their suitability for the railway environment, the suggested steps for further development.

2. CONVENTIONAL NDT METHOD

2.1 Ultrasonic Axle Testing

The railways of Great Britain have used ultrasonic axle testing techniques since the late 1940s. There are currently three principle scans that are used for the testing of axles:

- **Far end - looking at the entire axle length:** Will find cracks of approximately 5mm deep and larger
- **Near end - looking at the wheel bearing seat:** Will find cracks approximately 2mm deep and larger
- **High Angle - Looking at the wheel and gear seats:** Will find cracks approximately 1mm deep and larger

The main issue concerning the use of UAT, especially the far end scan, is a certain lack of confidence in finding cracks that may be present in view of the length of the axle being scanned. Also, to gain access to the axle ends where the probe must be placed, axle end covers must be removed together with bearing plates sometimes combined with WSP toothed wheels, as a result of which bearings may possibly be disturbed. If the removal and replacement of these components is not carefully controlled and carried out, bearing and wheel set failure could occur. The dangers are dirt ingress, and if the vehicle is moved (even very slightly – less than one turn) possible wheel bearing displacement.

The final issue with UAT is evaluation of the data, in the form of a trace where "noise" must be distinguished from the "signal". The evaluation is subjective and difficult to keep as an auditable record. Much rests in the skill and experience of the operator.

It has to be said that over the years, the British Railways, operators and the training process, has resulted in a highly skilled team of operators, but nevertheless, it is still a highly subjective process.

2.2 Magnetic Particle Inspection

Magnetic Particle Inspection (MPI) testing has been used since the 1930s, albeit not in railway applications, and has the ability to find very small cracks - less than 1mm in length. However, to find the crack all surface coatings must be removed from the axle (to apply the MPI solution and the magnetic field) and when a crack is identified, depending on the size of other cracks that may be present and their orientation, results will again be subjective. MPI gives no indication whether the crack is actually a scratch; i.e. MPI gives no indication of crack depth. As with UAT the results cannot be easily recorded for auditable records.

2.3 Dye Penetration

Very simply, dye penetration, like MPI, is a visual technique where the dye is applied to the test piece. It penetrates the crack giving a visual indication but like MPI the surface must be free of all surface coatings, results are again subjective and cannot easily be recorded for auditable purposes.

3. FRACTURE MECHANICS

It is very important to acknowledge fracture mechanics as the tool that predicts from a given crack size the crack propagation rate. For example, fracture mechanics principles would be used to determine if an axle could be safely used until the next wheelset overhaul, despite cracks being present. Fracture mechanics can only be used as a prediction tool for cracks of approximately 2mm deep or greater, depending on axle design (dimensions, materials and configuration). After this there is an almost linear relationship of crack growth with respect to time. To date there is no reliable method of predicting crack initiation and propagation of cracks less than 2mm in size. This is illustrated in the graph below.

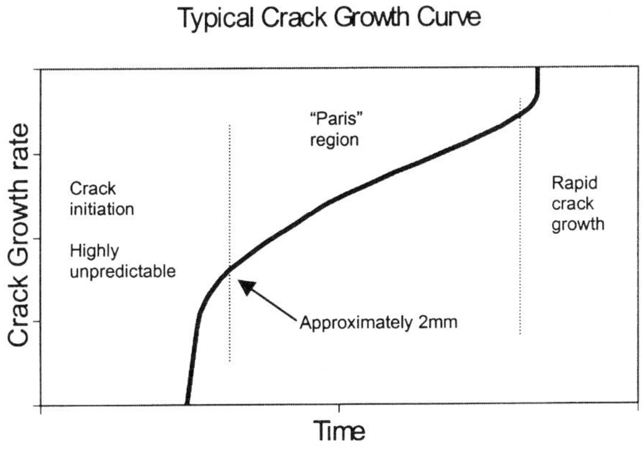

Typical Crack Growth Curve

Note the almost linear section used for predicting crack growth is referred to as the Paris Region. So, once a crack can of approximately 2mm in depth can be confidently detected with confirmation that it is a true crack and not a scratch/corrosion pit, it is possible to predict the safe working life of the wheelset.

4. THE NEW TECHNIQUES

The new techniques described appear to be very similar in use but are in fact rather different. However, they both exhibit the same very important features that represent a step change from the current conventional NDT methods, which are:

- The ability of near 100% confidence in finding cracks with a depth of 2mm and above
- The ability to find cracks without removing any surface coatings. Surface coatings can be up to approximately 10mm thick if non-metallic, and aluminium loaded paint and light corrosion do not affect the results
- The ability to size the crack in terms of length and depth differentiating between a scratch or corrosion from a true crack
- The advantage of providing results that are objective and can be stored in a database providing fully auditable records against wheelset serial numbers, that can be replayed at any time to show the testing method and technique

4.1 The Principles of the Techniques

Before going into detail about the two techniques outlined above, it is important to understand the detail of what the two techniques aim to achieve. Both techniques rely on inducing currents in the work piece that flow at 90° to the crack length. The currents may be ***uniform*** or ***non-uniform*** in distribution.

4.1.1 Analogy

- **Uniform induced currents**

  At this point it may be helpful to draw an analogy between a crack with electrical currents flowing around it, and a plank of wood placed in a stream of running water. As a stream flows, the uniform lines of running water approach the plank, and depending on the size of the plank, some of the water will flow around the ends, and some will flow underneath.

- **Non-uniform induced current**

  As for uniform induced current, but consider the plank of wood not in a stream of running water, but having jets of water fired on it. Water from jets near the end of the plank will flow around the end, and jets near the centre of the plank will flow both along the length of the plank and underneath it.

4.1.2 What both techniques aim to identify

It is the electric current flowing around the crack and underneath/along the crack that these techniques detect and measure. The flows are illustrated in the diagram below. Note the direction of the current flow around opposite ends of a crack that distinguish the start and end of a crack. For ease of illustration, the induced current streamlines are shown as ***uniform,*** i.e. equally spaced and of equal magnitude.

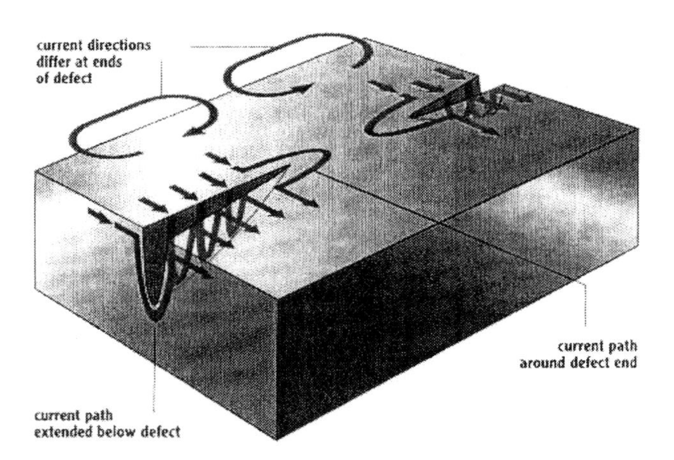

After measuring the current flows with pick up coils located above the work piece, in **fresh air** and through **surface coatings**, computer software is used to size the cracks. The difference in the two techniques is:

1. How they induce the current in the test piece, ***uniform*** or ***non-uniform.***
2. How the swirls around the end of the cracks are detected (to give a crack start and stop) for **crack length**.
3. How the current flows under/along the crack is detected to give a **crack depth**.

The next section in this paper describes how the two techniques work in detail.

4.2 Alternating Current Field Measurement (ACFM) by Technical Software Consultants (TSC)

In the simplest form ACFM utilises a probe design that has three coils.

- One coil to introduce a ***uniform*** alternating current flow within the test piece at 90° to the crack length.
- One coil whose axis is ***normal*** to the crack to measure the magnetic flux from the induced current flowing around the ends of the crack - this gives the start and the end of the crack, hence crack length.
- One coil whose axis is ***in line*** with the length of the crack to measure the magnetic flux from the induced current flowing under the crack - this gives crack depth.

Thus, having identified the start and the end of the crack along with the magnetic flux measured due to the current flow under the crack, computer software looks up those values from a database (based on a mathematical model of an elliptical crack subjected to a uniform current field), to give a crack depth.

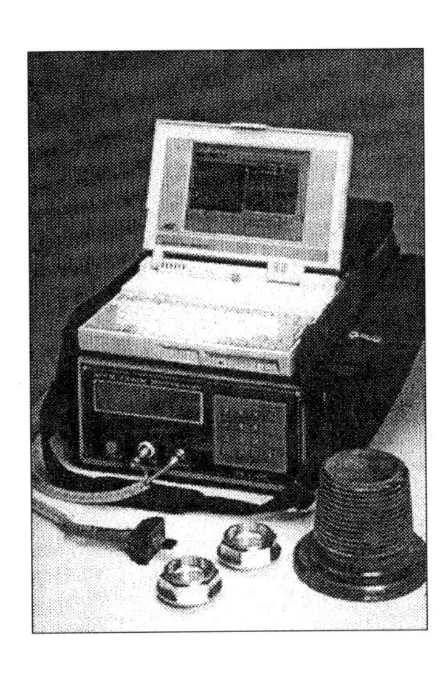

Developments and refinements to ACFM include having an array of coils to measure cracks so that the probe can be placed over a work piece without the need to physically move the probe, and where the scanning and the sizing is done electronically and controlled by software. Applications are wide and varied including use in the oil industry, theme parks, and aerospace. The ACFM technique is underwritten by Lloyds, and training can be carried out by The Welding Institute (TWI) to Personal Certification in NDT (PCN) levels.

4.3 Electro Magnetic Array (EMA) by ALSTOM/NEWT

In the simplest form the EMA probe has three main components.

- A wire through which an alternating current is passed to produce a ***non-uniform*** current flow within the test piece at 90º to the crack length.
- A coil the axis of which is ***normal*** to the test piece and directly over the wire (this induces the current flows) that detects the current flows in the work piece at the start and the end of the crack to give crack length.
- Another coil, the axis of which is also ***normal*** to the test piece, but offset from the first coil, and this measures the field gradient component of the current streamlines flowing along the length of the crack.

It is important to note at this stage that the induced current streamlines **are not uniform**. Going back to the analogy of the plank of wood in the stream where the current streamlines are uniform, non uniform streamlines can be analogised to firing a jet of water at the plank of wood and measuring how much of the jet is moving along the crack rather that under it. The field gradient of the decaying energy away from the centre of the induced current is measured to give crack depth.

A further analysis of the signal enables the lift off between the probe and test piece to be measured and recorded. This feature is used for quality control during the inspection.

Developments in EMA have led to solid probes that have several wires to induce the current and an array of coils to detect crack start/end and crack depth such that it is possible to electronically scan a work piece in all directions without moving the probe. EMA is also used by the oil industry, theme parks, and aerospace industries. The EMA technique is approved by Lloyds and other safety authorities. Training to Personal Certification in NDT (PCN) levels and ASNT is available via The Force Institute of Denmark (similar to TWI in the UK).

5 DEVELOPMENT OF NEW TECHNIQUES FOR THE RAIL INDUSTRY

It is important to remind ourselves that these techniques are principally designed to find surface cracks. In the case of axles with axle mounted components such as wheels, brake discs and gear wheels they cannot find cracks at the interfaces. In such cases of assembled wheels sets UAT will still have to be used. However there is an opportunity to combine the new techniques with the more accurate methods of UAT, such as high angle scans and near end scans which (see section 2.1) can reliably detect crack depths down to 2mm.

Developments to bring the new techniques into use in the railway environment of wheels and axles would include the following elements:

- Proving the new techniques are at least as good at crack detection as the current accepted techniques, such as MPI, including:
  - Probability of crack detection
  - Number of false detections
  - Auditing
- Approval of the techniques by a Vehicle Acceptance Body and acceptance by Group Standards.
- Approval of procedures by a Vehicle Acceptance Body and acceptance by Group Standards.
- Approval of operator training by a Vehicle Acceptance Body and acceptance by Group Standards.
- Validation of the claimed benefits with analysis of further benefits.

The aims of a development programme must be very clear. If, for example, the aim of the development was to gain confidence in crack detection such that fracture mechanics could be confidently applied to eliminate NDT testing between wheelset overhauls, then a high level development plan may look like:

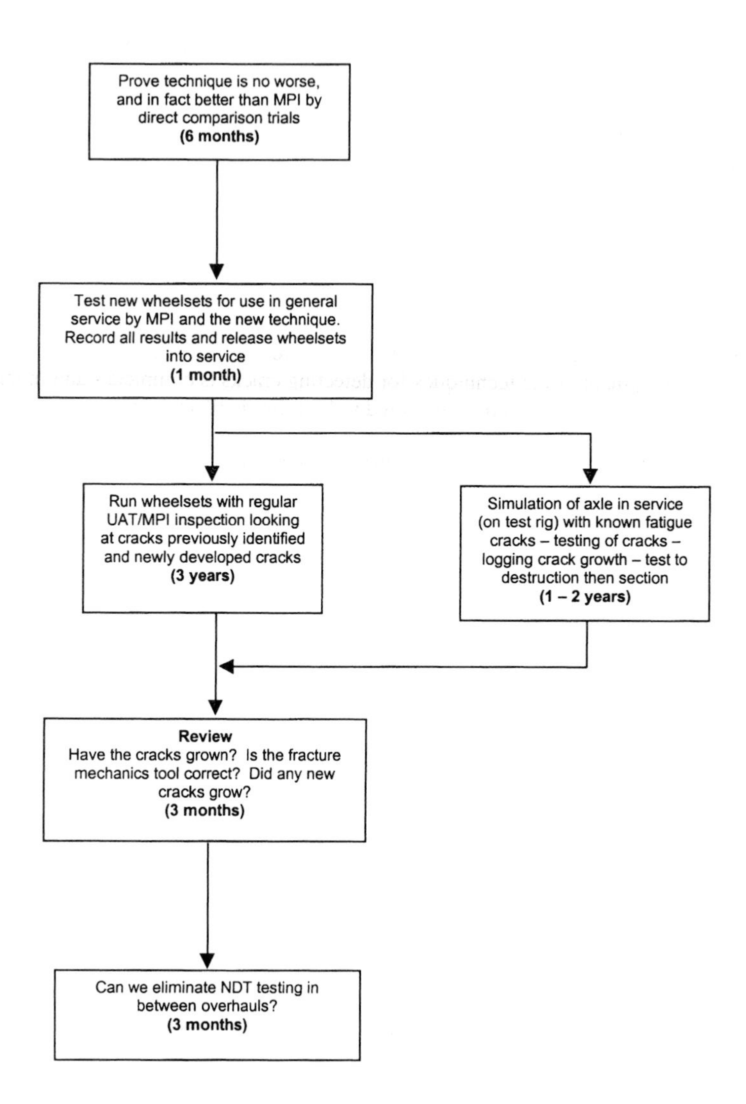
Prove technique is no worse, and in fact better than MPI by direct comparison trials
(6 months)
Test new wheelsets for use in general service by MPI and the new technique. Record all results and release wheelsets into service
(1 month)
Run wheelsets with regular UAT/MPI inspection looking at cracks previously identified and newly developed cracks
(3 years)
Simulation of axle in service (on test rig) with known fatigue cracks – testing of cracks – logging crack growth – test to destruction then section
(1 – 2 years)
Review
Have the cracks grown? Is the fracture mechanics tool correct? Did any new cracks grow?
(3 months)
Can we eliminate NDT testing in between overhauls?
(3 months)

6. CONCLUSION

ACFM and EMA present the potential for a step change improvement in NDT methods for railway wheels and axles to enhance the already excellent safety record of UK railway wheelsets.

ACFM and EMA are not just limited to the testing of axles, thus can be used for the NDT testing of any metallic railway component.

ACFM and EMA have the ability to reliably detect and fully size cracks. There is no need to remove surface coatings, and this could lead to optimised new axle coatings that never have to be removed, themselves offering significant advantages for axle life.

The new techniques produce objective and fully auditable results that can be electronically stored.

Development is required to package these techniques for the railway environment but the actual development of the techniques for detecting cracks is complete - and in this respect, the new techniques can be considered as "off the shelf" technology.

ACFM and EMA can become part of an automated process for crack detection.

Creative and imaginative applications could deliver real cost savings for the testing of axles, wheelset and any other metallic component in the rail industry.